SpringerBriefs in Agriculture

SpringerBriefs in Agriculture present concise summaries of cutting-edge research and practical applications across a wide spectrum of topics in agriculture with a fast turnaround time to publication. Featuring compact volumes of 50 to 125 pages, the series covers a range of content from professional to academic. Monographs of new material are considered for the SpringerBriefs in Agriculture series. Typical topics might include: A timely report of state-of-the art analytical techniques, a bridge between new research results, as published in journal articles, and a contextual literature review, a snapshot of a hot or emerging topic, an in-depth case study or technical example, a presentation of core concepts that students must understand in order to make independent contributions. Best practices or protocols to be followed. A series of short case studies/debates highlighting a specific angle.

Thomas Sinclair · Thomas W. Rufty

Bringing Skepticism to Crop Science

Springer

Thomas Sinclair
Crop and Soil Sciences Department
North Carolina State University
Raleigh, NC, USA

Thomas W. Rufty
Crop and Soil Sciences Department
North Carolina State University
Raleigh, NC, USA

ISSN 2211-808X   ISSN 2211-8098 (electronic)
SpringerBriefs in Agriculture
ISBN 978-3-031-14413-4   ISBN 978-3-031-14414-1 (eBook)
https://doi.org/10.1007/978-3-031-14414-1

This Springer imprint is published by the registered company Springer Nature Switzerland AG
The registered company address is: Gewerbestrasse 11, 6330 Cham, Switzerland

Contents

Chapter 1
The Role of Skepticism in Science

Skepticism—and its practitioners—is sometimes viewed unfavorably as being an unseemly approach to examining scientific research and to advancing scientific knowledge. This attitude may result in part from confusion between skepticism and cynicism. The cynic approaches research evaluations with a negative perspective assuming beforehand that research is ill-advised, done improperly, or poorly interpreted. Criticism from a 'grumpy' cynic is inevitable no matter how well your research is designed and executed. While the cynical attitude might sometimes result in useful criticisms, dealing with a cynic can be unpleasant, and even unproductive.

In contrast to the close-mindedness of a cynic, the skeptic approaches a topic with an open mind, suspending judgment until relevant evidence is available and evaluated. What was the novel approach in the evidence, or what was concerning? Are the research interpretations based on a full consideration of relevant evidence beyond the immediate evidence in reaching conclusions? Would it help to better define the context of the conclusions so that they are more focused on a narrower perspective, or alternatively, would a broader view be more helpful? The role of the skeptic is to openly evaluate research without prejudice or bias, and if appropriate, provide suggestions that may allow a more solid, evidence-supported conclusion. The goal of the skeptic should be to provide positive support that advances scientific understanding.

Skepticism as a questioning philosophy is not new. In Western culture, skepticism is traced to the Greek philosopher Pyrrho the Elis who lived from approximately 360 to 270 BCE. The word 'skeptic' comes from the Greek word meaning 'inquiring, reflective'. Embedded in the skeptic philosophy is the concept that nothing can be completely known with certainty. The skeptic is responsible to sort out with an open mind the possibility that a proposed idea is inconsistent with the evidence.

T. Sinclair and T. W. Rufty, *Bringing Skepticism to Crop Science*, SpringerBriefs in Agriculture, https://doi.org/10.1007/978-3-031-14414-1_1

PYRRHO.

1.1   Skepticism in Science

Given the philosophy of skepticism, there are obvious parallels with the formulation, testing, and analysis of scientific hypotheses, which are cornerstones of scientific inquiry. A hypothesis is a tentative explanation about a small segment, almost always a very small segment, of the cosmos and how it works. For the crop scientist, the usual scope of interest may be one of the many processes in the cropping system resulting in useful plant products. A hypothesis is not a prediction of the outcome of a study, but rather an explicit proposal explaining how a well-defined system 'works'.

A critical paradox of a hypothesis is that no matter how elegant it seems, a hypothesis cannot be proved true but only disproved. Any proposed proof of a hypothesis may be discredited by the next investigation based on a new perspective or a more sophisticated study. Therefore, much of scientific progress is the generation of evidence establishing that hypotheses are not true. And, even the disproof of a hypothesis needs to be considered skeptically due to the possibility of misleading experiments and interpretations resulting in mistaken 'disproof' (Kinrade and Denison, 2003).

If a hypothesis cannot be rejected, then the problem remains unresolved. A single test of a hypothesis, or even a number of tests, does not 'prove' a hypothesis. Consequently, the claim of 'validation' of an idea cannot be based on a limited set of tests. For this reason, the use of the word 'validated' is used much too readily in scientific writing. It takes many studies from differing perspectives, and usually a number of investigators, to promote a hypothesis to a 'validated theory'.

Skepticism is an essential feature of science. Objectivity and progress in scientific understanding require skepticism to move towards the 'truth'. Even then, new evidence and its skeptical evaluation may further alter or expand the truth. Just as Newtonian physics describes the everyday world, Einstein's theory of relativity offered an expanded description of the physical world. Almost surely, Einstein's theory of relativity will be given a new perspective by future research that attempts to develop a "universal equation of everything".

1.2 Skepticism is Not Easy

For scientists, however, an open, skeptical approach in interpreting evidence is not automatic or even easy. Human nature seems to be 'wired' to favor setting aside skepticism in favor of the less complicated, more superficial explanations. Why does the brain seem to deviate easily from skeptical analysis of evidence? In his book "Why People Believe Weird Things", Michael Shermer (1997) identified three reasons that contribute to answering this question.

- *Credo consolans*: "I believe because it is comforting/consoling". Scientific ideas almost never materialize without the investigator experiencing past frames of reference on the topic. Classroom instruction, input from colleagues, published papers, and even popular news impact the thinking of scientists. Conclusions that are counter to frameworks already established in the brain can be very unsettling. It is much more comfortable for scientists as human beings to draw conclusions that are consistent with the established framework. It feels good to present ideas that are in some way compatible with previous conclusions, even if the rigors of skepticism are set aside.
- Immediate gratification: Ambiguity is generally unsettling to the human brain. Rather than leave things unresolved, it feels better to have an explanation for the evidence that was collected, even if the evidence is somewhat ambiguous. A rapid resolution of a problem is appealing. Sifting through evidence beyond that which the scientist collects takes time and effort, and the time and effort expand when the focus on the topic is broadened. Skepticism often does not lead to immediate gratification.
- Simplicity: The world is complex and complicated, especially the biological world. Resolution of complexity is extremely challenging and it is much quicker and easier to propose a simple answer. This is especially true in view of Einstein's urging that "Everything should be made as simple as possible, but no simpler."

The first part of the quotation is easily remembered in developing simple ideas. It is the duty of the skeptic to remember the last phrase "but no simpler"!

This book examines the need for skepticism in thinking about some of the prevalent ideas in crop science, particularly those that potentially contribute to crop yield increase. Specifically, seven topics are reviewed from a skeptics perspective: photosynthesis, seed number, nitrogen use efficiency, osmolyte accumulation, water use efficiency, crop transpiration prediction, and unconfirmed field observations. All of these topics have essential relevance in understanding the interaction between crop production and the environment. Importantly, these understandings are crucial in applying conclusions from crop science research to two of the 'big' issues facing the world: the needs for increases in environmentally secure food production and for mitigation of the challenges of climate change. Without skepticism, solutions proposed for these global issues may well be illusions.

References

Kinraide TB, Denison RF (2003) Strong inference: the way of science. Am. Bio. Teach. 65:419–424
Shermer M (1997) Why people believe weird things. WH Freeman Co., New York

Chapter 2
Cautionary Alerts for Skeptics

As, pointed out in the previous chapter, skepticism is a hallmark of solid science. This is especially true in crop science where a large number of variables impact experiments, and many of the environmental and biological variables can confound interpretation of the results. It is more than prudent to suggest conclusions from almost all crop science studies are 'tentative'. It is virtually certain that results and conclusions could differ when tested in a new environment with a different species or genotype. The understanding of crop science seems to be almost always destined to be essentially based on a set of tentative hypotheses.

Given that there is a high degree of uncertainty in crop science, general "caution alerts" can be helpful for the skeptic when evaluating new evidence and hypotheses. In this chapter, four alerts of caution in evaluating evidence are suggested as being particularly relevant in crop science. The alerts are reliance on correlation analysis, use of phenomenological descriptions, ignoring differential time scales, and inadequate field tests. While these warnings are not new, it seems in many cases they have been readily ignored both in the original published sources evaluating a hypothesis and by those that have accepted a hypothesis.

2.1 Correlation

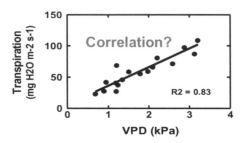

A large amount of analytical effort applied to crop science data is often devoted to examining the correlation between two variables. In a correlation analysis done by regression, it is implicitly assumed that one variable is 'independent' and the other variable(s) is assumed to be 'dependent' on the first variable. Without full awareness of this assumed relationship between variables, cause-and-effect relationship can be implicitly concluded between the two variables. This implication runs the risk of exposing the investigator's bias about how the topic is to be structured. The skeptic needs to be aware that it may not be at all clear which variable is independent and which is the dependent variable. Or, even if two variables are found to be statistically correlated, the correlation in itself does not prove a causal relationship.

There are many examples of correlation that do not necessarily indicate a direct cause-and-effect relationship. Being a 'morning person' myself, one possible example might result from data on the early morning activity of birds and the disappearance of dew off of vegetation. Observations could be taken over time and the data subjected to correlation analysis. While I am not sure such data exists, it seems highly likely that as the morning progresses through sunrise that birds will become more active and the dew disappears. If bird activity is assumed the independent variable, should it be concluded that somehow wing flapping by flying birds was the 'cause' for the 'effect' of the disappearance of dew? Did the flapping of wings stir the air so that dew evaporated? Of course, this 'analysis' is silly and there is no direct relationship between the two events other than they are both linked to the rising sun. A correlation might exist in this bird-dew example, but a cause-and-effect relationship does not.

The caution for crop science is that there are many interacting processes in play and correlation should not be readily assumed to be causal. As with any hypothesis, the establishment of causal relationships require multiple sources of evidence obtained from various perspectives. In crop science, correlations should not be accepted as explanatory without being subjected to the necessary skeptical challenge that explores various options to account for the correlative relationship. As discussed later, some major hypotheses and conclusions in crop science are based on correlations with little supporting evidence concerning any possible, direct mechanistic linkage.

2.2 Phenomenological Descriptions

"Phenomenology" is a qualitative approach coming from social sciences to describe or interpret phenomena. The rigor of phenomenological descriptions has been proposed to be somewhat flexible under the term "bridling" (Dahlberg 2006; Vagle et al. 2009). "Bridling" comes from the experience of riding horses when sometimes the reins are held loosely and sometimes held tightly. Therefore, a range in the level of qualitative description can vary according to the phenomenon being described.

In crop science, however, purely qualitative descriptions of phenomena are generally felt to be incomplete so quantitative relationships are sought to describe phenomena. Such 'bridging' into quantitative expressions occurs even though some of the components of the relationship may not be fully understood or are described incompletely. A 'placeholder' variable may be included in a phenomenological expression to complete the equation although understanding of the placeholder term requires further scrutiny. A major challenge for the scientist is to discern whether an expression is a phenomenological relationship with its embedded ambiguity, or is it truly a derived equation based on fundamental mechanistic relationships.

There have been cases in crop science where phenomenological descriptions have become prevalent and the phenomenological nature of the description is not widely recognized. In some cases, placeholder variables in phenomenological expressions have themselves become major focuses of research. One such example is in the analysis of water use efficiency, which is discussed in Chap. 7. The argument in that chapter is that it is necessary to look beyond a phenomenological expression, if possible, and pursue a more rigorous expression based on clearly defined mechanisms.

2.3   Time Insensitivity

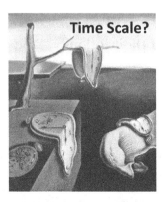

The Persistence of Memory by Salvador Dali

Processes that control crop yield are often described at the basic process level based on an instantaneous, or nearly instantaneous time scale. These basic descriptions are then applied to crop performance at time scales of a day, a week, or even the entire growing season. A major challenge that is often overlooked is to appropriately and accurately integrate instantaneous descriptions of processes to represent the process over increasing time scales. How can instantaneous processes be described so that they accurately reflect integration over larger times scales?

One common approach to deal with longer-time processes has been simply to ignore the nature of the variability over time and simply assume substitute representative values. An example is presented in Chap. 8 in which an instantaneous energy balance description of crop canopy transpiration is applied to longer time periods using various approximations for some of the input values. The input values may be based on unsupported assumptions, or simply 'guesstimated'. As a result, there can be a high degree of uncertainty in the calculated transpiration values.

2.4   Inadequate Field Tests

While the need for rigorous field testing is widely recognized, too often data sources are applied in scientific publications that have not met accepted standards for meaningful yield data (Sinclair and Cassman 2004). One of the egregious sources of inadequate yield results are from farmers' yield-contest winners as reviewed in Chap. 9. The results from farmers' yield contests are referenced in scientific analysis even though such contest results do not meet scientific standards. The farmers' fields from which the yields are reported are not replicated and supporting evidence about the nature of the environment in which the yields were achieved are not available. Also, and perhaps the most worrisome aspect of yield contest reports, is that the harvest data are vulnerable either to self-deception or outright intentional deception. Consequently, on occasion yields of contest winners substantially exceed scientifically generated research data, or are even greater than those possible given the resources available for crop growth.

Another source of inappropriate crop yield results are presented in papers that include supporting evidence about the yields of genetically transformed plants. These data are presented as evidence of improved crop germplasm, and in some cases, it is concluded the results are evidence of a solution for improving global food supply. Very rarely do any of these yield studies meet the accepted standards for agronomic evaluation. Were plants grown in insufficiently large field plots to allow harvest areas that were well bordered? Were plants grown at appropriate densities when compared to commercial situations? Were there sufficient plot replications to allow useful statistical analysis? And importantly, were the tests done in a sufficient number of environments to support conclusions about yield response?

A final caution in interpreting the relevance of yield results from genetic transformation studies is the nature of the germplasm to which yield increases are compared. The reference germplasm is often the 'wild type' from which the transgenic material was derived. As discussed in Chap. 9, such comparisons do not allow interpretations about actual crop yield improvement. The much more important issue is whether the transformed germplasm has superior yields compared to existing commercial germplasm grown using relevant agronomic practices.

References

Dahlberg K (2006) The essence of essences—the search for meaning structures in phenomenological analysis of life world phenomena. Int J Qual Studies Health Well-Being 1:11–19

Sinclair TR, Cassman KG (2004) Agronomic UFOs. Field Crops Res 88:9–10

Vagle MD, Hughes HE, Durbin DJ (2009) Remaining skeptical: bridling for and with one another. Field Meth 21:347–367

Chapter 3
Photosynthesis and Yield

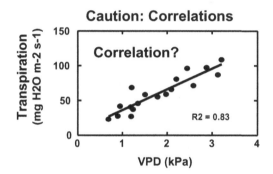

The logic for proposing that crop yield can be increased by increasing leaf photosynthesis rate seems straightforward. The mass of plants is built on the 'muscle' of carbon molecules, with elemental carbon accounting for about 40% of plant dry weight. Thus, increasing crop yield necessarily requires more carbon and seemingly could be driven by increased input of carbon, i.e. increased leaf photosynthesis rates. This view is offered either implicitly (Araus et al. 2021) or explicitly (Long et al. 2006; Simkin et al. 2019) as an underlying basis for considerable photosynthesis research purported to be a means to increase crop yields.

In actuality, there are only a few reports directly supporting this hypothesis and these are based on **correlations** between yield and measurements of leaf photosynthesis rate under specific circumstances (Sun et al. 2017; Gleason et al. 2021). As with all hypotheses, the proposal that enhanced photosynthesis rate can increase crop yields must be examined with skepticism. A review of studies involving field evaluations studying genetically diverse populations within crop species have shown overall a very weak link or no correlations between yield and leaf photosynthesis rates (Sinclair et al. 2019). Recently, this view was supported in the review by Araus et al. (2021) who concluded that "there is substantial literature spanning several decades that has shown no correlation between grain yield and the photosynthetic rate of leaves." Passioura (2020) summed up the situation by pointing out that "The many attempts to increase PY [potential yield] by manipulating the biochemistry and physiology of photosynthesis and respiration have so far been unsuccessful." A recent search of CAB Abstracts (14 Dec 2021) identified nearly 890 papers that had both "photosynthesis" and "yield" in their title, but there appears to be no commercial cultivars that were released based on increased photosynthesis rates associated with yield increase.

3.1 Leaf vs. Canopy Carbon Accumulation

There are numerous physiological reasons why the apparently straightforward logic of increasing carbon inputs into plants does not translate into increased crop yield. One reason for the failure in the apparent logic is that photosynthesis capacity of individual leaves when combined does not linearly translate into canopy carbon assimilation rate. That is, there is a diminishing return in canopy assimilation rate with increasing leaf carbon exchange rate (CER). This relationship was demonstrated in the theoretical derivation for canopy growth expressed as radiation use efficiency (Monteith 1977; Sinclair and Horie 1989) as shown in Fig. 3.1. RUE is defined as crop mass accumulation per unit of radiation intercepted by a crop canopy.

It is important to recognize that the results presented in Fig. 3.1 are based on a mechanistic description of canopy gas exchange although several simplifying assumptions were required in the derivation (Sinclair and Horie 1989). The key assumptions were: (1) leaf segments in the canopy at any instant are subjected to either direct solar radiation or shaded conditions; (2) leaf area and average intercepted radiation flux density of each leaf type are calculated based on a horizontally random distribution of leaves; (3) an estimate of leaf photosynthesis rate for sunlit and shaded leaf segments calculated from the average radiation flux density on each leaf type; an asymptotic exponential function accurately describes the photosynthetic light-response curve (Boote and Jones 1987) with the equation scaled to the specific

Fig. 3.1 Graph of radiation use efficiency derived from light-saturate leaf carbon exchange rate (CER_{max}) for cereals and soybean (Sinclair and Horie 1989)

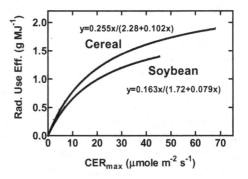

light-saturated photosynthesis rate (CER_{max}) for each species; (4) growth respiration rate is defined by the biochemical composition (i.e. carbohydrate, protein, and lipid) of the synthesized plant products of each crop species (Penning de Vries 1975); and (5) maintenance respiration rate is assumed to be approximately equal to 15% of hexose synthesis.

The two critical variables in the derived radiation use efficiency that account for differences among species are the energy content of the plant mass and CER_{max}. The range in seed mass synthesized based on energy content of various crop species per unit hexose input is from 0.42 g g^{-1} for sesame (*Sesamum indicum* L.) to 0.75 g g^{-1} for rice (*Oryza sativa* L.) and barley (*Hordeum vulgare* L.) as shown as the abscissa variable in Fig. 3.2. The impact of the energy content on RUE among species is illustrated in Fig. 3.1. Differences in the RUE curves result between cereals (plant mass of comparatively low energy content) and soybean (*Glycine max* Merr. L.) (plant mass of comparatively high energy content).

The expected RUE values for cultivars of commercial crop species can be obtained from the results presented in Fig. 3.1 using field observations of leaf CER. For soybean, leaf CER measured for several cultivars over at least 20 consecutive days was reported to reach 40 μmole m^{-2} s^{-1} and above (Sinclair 1980). For well-fertilized maize (*Zea mays* L.), leaf CER has been found to be at least 50 μmole m^{-2} s^{-1} (Muchow and Sinclair 1994). Radiation use efficiency predicted from the observed CER values and the equations in Fig. 3.1 are 1.33 g MJ^{-1} for soybean and 1.73 g MJ^{-1} for maize. These radiation use efficiency estimates are only slightly less than the reported top-end, experimental observations of 1.42 g MJ^{-1} for soybean

Fig. 3.2 Graph of nitrogen required per unit of seed mass vs. the seed mass synthesized per unit of photosynthate supplied for seed growth (Sinclair and de Wit 1975)

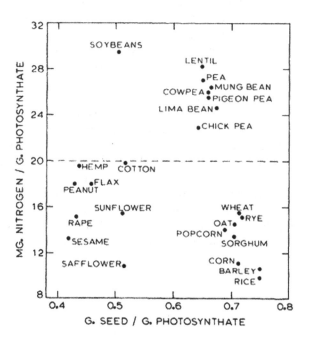

(Purcell et al. 2002) and 1.86 g MJ^{-1} for maize (Lindquist et al. 2005). Expectation of radiation use efficiency values substantially greater than these predicted and observed values appear unrealistic given the nature of the radiation use efficiency response to increasing CER.

3.2   Seed Yield

Examination of the synthesis hierarchy going from the molecular level to seed yield demonstrates further the challenge of increasing crop yield by increasing the molecular capacity for photosynthetic activity. Such a set of calculations were presented by Sinclair (2004) for soybean by using experimentally measured sensitivities of each step in moving through the hierarchy to seed yield. Figure 3.3 shows the series of calculations starting with an assumed technological advance giving a 50% increase in Rubisco RNA abundance. This assumed 50% increase eventually was calculated to result in an increase of only 18% in crop mass accumulation, which is equivalent to RUE increase. The critical, final step in the hierarchy is the calculation of seed mass. Based on carbon partitioning in the plant and the energy requirements for seed synthesis, the soybean seed yield increase in this example is calculated to be only 6% with the optimistic assumption that the plant was able to also accumulate the additional nitrogen required by the increased seed mass. However, if the plant was not able to accumulate additional nitrogen, the partitioning of nitrogen to the early-season increase in vegetative tissue mass would result in less nitrogen ultimately available for seed growth. In this case of less nitrogen for seed growth, the yield is predicted to decrease(!) by 6% as a result of the 50% increased molecular capacity for leaf photosynthesis. The experimental observations of Mitchel et al. (1993) under a CO_2 enriched atmosphere showed exactly such a response in decreased grain yield in wheat (*Triticum aestivum* L.) associated with increase in CO_2 assimilation in a treatment of low nitrogen availability.

The results in Fig. 3.3, highlight the critical role of nitrogen in determining ultimate crop yield. That is, seed development is not simply limited by rates of carbon accumulation but has a quantitative requirement for nutrients, especially nitrogen. As with all growth processes throughout the plant, nitrogen is required for sustained cell division, and to produce the proteins that are characteristic components of seeds. While carbon is a partner in yield generation, the acquisition of nitrogen and its role in all regulatory processes in the plant causes it to be often the 'regulator' of crop growth. A quantitative pairing of nitrogen and carbon accumulation (Fig. 3.4) is essential in the growth of new tissues, including seeds.

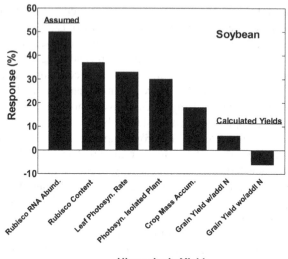

Hierarchy in Yield

Fig. 3.3 Hierarchy is estimating seed yield of soybean (Sinclair et al. 2004) starting with an assumed increase in Rubisco RNA abundance of 50%. The steps in the calculation were based on experimental results for RNA abundance to Rubisco concentration (Jiang et al. 1993), Rubisco concentration to leaf photosynthesis rate (Jiang et al. 1993), leaf photosynthesis rate to plant photosynthesis (Sinclair et al. 2004), and plant photosynthesis to crop mass accumulation (based on results in Fig. 3.1). The final step (Sinclair et al. 2004) shows two yield options based on the assumption of either with additional nitrogen accumulation to match seed requirements (6% yield increase) or without additional nitrogen accumulation by the crop (6% seed yield decrease)

Fig. 3.4 Diagram of requirement for nitrogen and carbon in the synthesis of seeds (Sinclair et al. 2019)

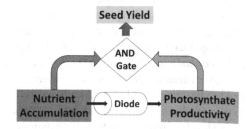

3.3 Nitrogen and Seed Yield

The nitrogen limitation in achieving high yield can be described using the following simple phenomenological equation based on nitrogen accumulation in seeds.

$$Y = NHI \cdot N/GN \tag{3.1}$$

where NHI = harvest index, N = amount of nitrogen accumulated by the plant, and GN = grain nitrogen concentration. The critical feature of Eq. (3.1) is that to achieve high grain yield large amounts of N must be accumulated in seeds.

The essential role of nitrogen is determining crop yield is clearly shown by historical records. Natural fixed nitrogen input to the soil is commonly something in the range of 20–30 kg N ha^{-1} y^{-1}. These levels of nitrogen input and NHI = 0.8 and GN = 21 mg N g^{-1} when used in Eq. 3.1, give annual yield estimates of only 760 to 1140 kg ha^{-1}, respectively. Understandably, through most of human history cereal grain yields in the absence of nitrogen application to fields in some form were only about 1000 g m^{-2} or less (Sinclair and Sinclair 2010). It was only when manufactured nitrogen fertilizer became widely available after World War II that the Green Revolution with major increases in crop yields began. As pointed out by Norman Borlaug (1972), "chemical fertilizer is the fuel that has powered the [Green Revolution] forward thrust."

As an aside, plant breeding was not the original basis for yield increase. Double-cross maize hybrids were commercialized in the 1930s and 1940s, but these hybrids did not result in notable increases in yield. The advantage of the early maize hybrids was that they produced uniform plants, which facilitated the advent of mechanized crop harvesting.

Equation 3.1 can be rearranged and then used to calculate the large amounts of nitrogen that must be accumulated for grain production. To achieve a maize (*Zea mays* L.) grain dry weight yield of 16,000 kg ha^{-1} (16 t ha^{-1}), GN = 12.2 mg N g^{-1}, and NHI = 0.80 (Parco et al. 2020), rearranged Eq. (3.1) results in an estimate of crop accumulated nitrogen of 260 kg ha^{-1}. For a wheat grain dry weight yield of 10,000 kg ha^{-1}, GN = 21 mg N g^{-1}, and NHI = 0.80 (da Silva et al. 2014), the requirement for accumulated nitrogen is equal to 262.5 kg ha^{-1}.

While the required amount of accumulated nitrogen is large, at first glance it seems practical. However, denitrification and leaching of nitrogen applied to the soil results in nitrogen having only an ephemeral existence in the soil. Typically, only about 40% or less of applied nitrogen fertilizer is recovered by a growing crop. Therefore, for a crop to accumulate 260 kg N ha^{-1} the original availability of nitrogen in the soil may need to be something like 650 kg N ha^{-1}! This very high level of nitrogen addition to the soil also means that nearly 400 kg N ha^{-1} will be released to the environment. These levels of nitrogen present both major economic and environmental challenges.

In addition to nitrogen's synthetic role in seed formation described above, the market value of the grain is sensitive to seed nitrogen content expressed as protein concentration. For most major crop species, the price of grain can be severely discounted if the seed protein concentration is less than various thresholds. For example, in wheat the marketplace seeks grain protein concentrations of 13% (i.e. 21 mg N g^{-1}) as required for wheat flour in the baking industry. Only maize among the major crop species has been allowed to have decreasing nitrogen concentration to achieve increasing yields. The nitrogen concentration in maize grain has now decreased from about 13.5 to about 12.2 mg N g^{-1} over the past 30 years (Scott et al. 2006; DeBruin et al. 2017) allowing about a 10% yield increase.

If the large amounts of nitrogen estimated from Eq. 3.1 are not accumulated in the plant, what is the impact on carbon accumulation? During vegetative development, low nitrogen levels in the plant can result in decreased leaf area development or early leaf abscission. Loss in leaf area due to limited nitrogen in the plant can result in a decrease in light interception by the crop canopy. In addition, low levels of accumulated nitrogen can result in decreased leaf nitrogen concentrations, which in turn result in decreased carbon exchange rates. As discussed previously and shown in Fig. 3.1, low carbon exchange rates lead directly to low leaf photosynthesis rates.

During seed growth, accumulated carbon must be matched with nitrogen in the process of growing new seed mass (Fig. 3.4). Under conditions of low nitrogen availability in plants, there may be an accumulation of excess carbon compounds in the plant. The accumulation of carbon has the potential to result in a negative feedback on photosynthetic rate (Brinkert et al. 2016; Heyneke and Fernie 2018; Marquardt et al. 2021). Timm (2020) suggested that photorespiration can have a beneficial role in this situation by moderating carbon accumulation in leaves. Additional means to minimize carbon accumulation in the plants are mechanisms to dispose of excess carbon during plant development including exudation and leaching of carbon compounds from plant tissues (Thomas and Sadras 2001).

Increasing crop yields to levels substantially greater than those currently harvested can clearly be limited by nitrogen availability in the soil. It is not a simple task to effectively balance the challenges of applying large amount of nitrogen fertilizer, minimizing environmental-nitrogen pollution, assuring temporal patterns of nitrogen availability to match crop nitrogen uptake needs, and finally, obtaining the required transfer of nitrogen from vegetative tissues to growing seeds. For the skeptic, all the challenges in regards to crop nitrogen accumulation must be resolved before crop yield is actually limited by carbon accumulation. Water and phosphorus availability are two additional limitations that need to be resolved before potential carbon accumulation is limiting of crop yield.

References

Araus JL, Sanchez-Bragado R, Vicente R (2021) Improving crop yield and resilience through optimization of photosynthesis panacea or pipe dream? J Exp Bot 72:3836–3955

Boote KJ, Jones JW (1987) Equations to define canopy photosynthesis from given quantum efficiency, maximum leaf rate, light extinction, leaf area index, and photon flux density. Progress in Photosyn Res 4:415–418

Borlaug NE (1972) The Green Revolution, Peace and Humanity. CIMMYT Reprints and Translation Series No. 3, International Maize and Wheat Improvement Center

Brinkert K, De Causmaecker S, Krieger-Liszkay A, Fantuzzi A, Rutherford AW (2016) Bicarbonate-induced redox tuning in photosystem II for regulation and protection. Proc Nat Acad Sci. 113:12144–12149

Da Silva CL, Benin G, Bornhofen E, Todeschini MH, Dallo SC, Sassi LHS (2014) Characterization of Brazilian wheat cultivars in terms of nitrogen use efficiency. Bragantia Campinas 73:87–96

DeBruin JL, Schusssler JR, Mo SH, Cooper M (2017) Grain yield and nitrogen accumulation in maize hybrids released during 1934 to 2013 in the US Midwest. Crop Sci 57:1431–1446

Gleaseon SM, Nalezny L, Hunter C, Bensen R, Chintamanani S, Comas LN (2021) Growth and grain yild of eight maize hybrids are aligned with water transport, stomatal conductance, and photosynthesis in semi-arid irrigated system. Physiol Plant 172:1941–1949

Heyneke E, Fernie AR (2018) Metabolic regulation of photosynthesis. Biochem Soc Trans 46:321–328

Jiang CZ, Rodermel SR, Shibels RM (1993) Photosynthesis, rubisco activity and amount, and their regulation by transcription in senescing soybean leaves. Plant Physiol 101:105–112

Long SP, Zhu XG, Naidu SL, Ort DR (2006) Can improvement in photosynthesis increase crop yields? Plant Cell Env 29:315–330

Lindquist JL, Arkebauer TJ, Walters DT, Cassman KG, Dobermann A (2005) Maize radiation use efficiency under optimal growth conditions. Agron J 97:72–78

Marquardt A, Henry RJ, Botha FC (2021) Effect of sugar feedback regulation on major genes and proteins of photosynthesis in sugarcane leaves. Plant Physiol Biochem 158:321–333

Mitchell RAC, Mitchell VJ, Driscoll SP, Franklin J, Lawlor DW (1993) Effects of increased CO_2 concentration and temperature on growth and yield of winter wheat at two levels of nitrogen application. Plant Cell Envir 16:521–529

Monteith JL (1977) Climate and its efficiency of crop production in Britain. Trans R Soc London B 281:277–294

Muchow RC, Sinclair TR (1994) Nitrogen response of leaf photosynthesis and canopy radiation use efficiency in field-grown maize and sorghum. Crop Sci 34:721–727

Parco M, Ciampitti IA, D'Andrea KE, Maddonni GA (2020) Prolificacy and nitrogen internal efficiency in maize crops. Field Crops Res 256:107912

Passioura JB (2020) Translational research in agriculture. Can we do it better? Crop Pasture Sci 71:517–528

Penning de Vries, FWT (1975) Use of assimilates in higher plants. In: Photosynthesis and Productivity in Different Environments , Ed: Cooper J. Cambridge University Press, p 454–480

Purcell LC, Ball RA, Reaper JK III, Vories ED (2002) Radiation use efficiency and biomass production in soybean at different plant population densities. Crop Sci 42:172–177

Scott MP, Edwards JW, Bell CP, Schussler JR, Smith JS (2006) Grain composition and amino acid content in maize cultivars representing 80 years of commercial maize varieties. Maydica 51:417–423

Simkin AJ, Lopez-Calcagno PE, Raines CA (2019) Feeding the world: improving photosynthetic efficiency for sustainable crop production. J Exp Bot 70:1119–1140

Sinclair TR (1980) Leaf CER from post-flowering to senescence of field-grown soybean cultivars. Crop Sci 20:196–200

Sinclair TR, de Wit CT (1975) Photosynthate and nitrogen requirements for seed production by various crops. Science 189:565–567

Sinclair TR, Horie T (1989) Leaf nitrogen, photosynthesis, and crop radiation use efficiency. Rev Crop Sci 29:90–98

Sinclair TR, Sinclair CJ (2010) Bread, beer & the seeds of change: agriculture's imprint on world history. CABI, Wallingford, UK

Sinclair TR, Purcell LC, Sneller CH (2004) Crop transformation and the challenge to increase yield potential. Trends Plant Sci 9:70–75

Sinclair TR, Rufty TW, Lewis RS (2019) Increasing photosynthesis: unlikely solution for world food problem. Trends Plant Sci 24:1032–1039

Sun Y, Yan X, Zhang S, Want N (2017) Grain yield and associated photosynthesis characteristics during dryland winter wheat cultivar replacement since 1940 on the Loess Plateau as affected by seeding rate. J Food Agric 29:51–68

Thomas H, Sadras VO (2001) The capture and gratuitous disposal of resources by plants. Func Ecol 15:3–12

Timm S (2020) The impact of photorespiration on plant primary metabolism through metabolic and redox regulation. Biochem Soc Trans 48:2495–2504

Chapter 4
Increasing Seed Number

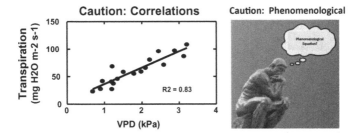

Caution: Correlations — Caution: Phenomenological

There are literally hundreds of papers in which crop yield and seed number are reported showing that these two variables are **correlated**. A simple **phenomenological equation** is used to describe seed yield (Y) as a function of seed number (SN) multiplied by average seed mass (SM). That is,

$$Y = SN \cdot SM \tag{4.1}$$

In some cases, the SN term is expanded to include additional reproductive components such as spikelet number and floret number (Slafer 2003). Since SM is often observed to be within a fairly stable range, the **phenomenological expression** is reduced to argue that Y is functionally dependent on SN.

To illustrate the correlation expression, Fig. 4.1 shows experimental results of yield graphed against seed number for soybean (*Glycine max* L. Merr.) cultivars grown in Argentina and US (Rotundo et al. 2012). A linear correlation was found in both countries between seed yield and seed number over a fairly wide range of seed number. The fact that yield is expressed as being dependent on seed number implies a causal dependence. That is, crop yield is presented as being directly dependent on the number of seeds produced by the plant. For instance, Abbate et al. (1995) concluded that "the number of grains m^{-2} is the most important yield determinant in wheat" based on the high correlation between yield and SN. Slafer (2003) suggested

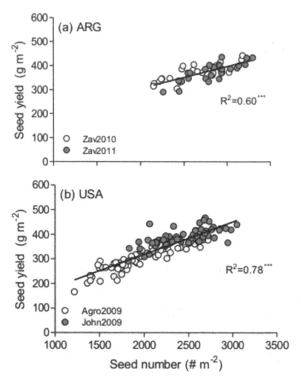

Fig. 4.1 Graph of soybean seed yield against seed number for crops grown in **a** Argentina (ARG) and in **b** the United States (USA) (Rotundo et al. 2012)

that "If we can increase the number of grains per unit land area, then yield would be significantly increased."

4.1 Resource Availability and Yield

Yet, the above claims about yield dependence on seed number were made without detailed studies to understand the basic physiological processes controlling seed set and grain number. In reviews of the relationship between yield and SN, Sinclair and Jamieson (2006, 2008) concluded that resource availability was critical in determining both yield and SN. While there can be a correlation between yield and SN, it should not be assumed this is a causal relationship. The relationship is somewhat like the example given in Chap. 2 of the possible correlation between morning activity of birds and dew evaporation. While in the bird-dew example the two variables are likely correlated to some degree, they are linked only by the common dependence on the rising sun.

In a study of soybean and maize (*Zea mays* L.), Egli (2019) explored the relationship between seed number and the crop growth rate, which was taken as an expression of resource availability. He found the two variables were correlated ($r^2 =$ 0.28 and 0.43 for maize and soybean, respectively). The inference Egli made from the analysis was that crop growth and seed number were both a result of resource input. Sinclair and Jamieson (2006, 2008) reached their conclusion in regards to wheat (*Triticum aestivum* L.) based on earlier studies in which resource availability was the common factor that was limiting both SN and yield. The fact that both SN and yield are dependent on resource availability in the plant means the two terms would be expected to be correlated, but again the correlation should not be interpreted as indicating causality.

In the sequence of events leading to grain set, almost always there is a large abundance of ovules produced, many of which are discarded both before and after fertilization. For wheat, only 1/3 or less of the florets were found to ultimately produce grain (Sibony and Pinthus 1988). A critical question is what causes only a fraction of the ovules to develop into grain. A hypothesis to explain the failure of wheat florets from developing into grain is carbohydrate deprivation. This hypothesis is based on the idea that inadequate carbon in the developing florets results in their abortion. Shading experiments with wheat have been offered as support for the hypothesis (Savin and Slafer 1991; Yang et al. 2020), but shading impacts many processes in the plant. One factor not impacted by shading was the soluble carbohydrate concentration in wheat spikelets (Mishra and Mohaptra 1987). They found spikelet soluble carbonhydrate concentration varied little prior to and through anthesis even though differences in SN and yield were obtained Such contrary evidence showing no variation in the availability of photosynthate underscores the need for caution in invoking the photosynthate-deprivation hypothesis as the primary cause of seed set failure.

Further, an implicit assumption of the carbohydrate-deprivation hypothesis is that increased seed set would result in stimulated crop carbon assimilation to support the greater seed growth rates resulting from greater seed numbers. However, there appears to be no evidence that leaf photosynthesis rates of wheat leaves were increased in the progression during reproductive development starting before anthesis and going into grain fill regardless of the ultimate seed number eventually produced by plants (Gargcia et al. 1998; Reynolds et al. 2005).

A corollary to the carbohydrate-deprivation hypothesis is that nitrogen supply in wheat to developing florets and the consequent setting of seeds may be linked to the ultimate SN (Sinclair and Jamieson 2006, 2008). While this hypothesis is not proven, it is consistent with the results of several studies showing a close relationship between the amount of nitrogen accumulated in the wheat spike at anthesis and SN (Abbate et al. 1995; Demotes-Mainard et al. 1999). These results support the fact that developing embryos require substantial nitrogen in the synthesis of their biochemical constituents and as components of the seed. So, ultimately SN is related to a large extent on the amount of nitrogen being supplied to and accumulating in the spikelets.

4.2   Component Compensation

An additional issue in interpreting the relationship between SN and yield is the high level of negative correlation between SN and SM in many crop species (Adams 1967). That is, smaller seed numbers were associated with larger seeds. Adams labelled this negative association as 'component compensation'. He attributed component compensation to the "nutrients and metabolites needed in the initiation and elaboration of reproductive structures". That is, a range of SN and SM could develop in plants, but the compensation between SN and SM inevitably resulted in similar total seed mass among plants of the same size.

A study by Spaeth and Sinclair (1984) illustrated the extent of component compensation in soybean. In their study, all plants in a length of row were harvested, so that a plant population with a wide range of individual plant masses was collected. For each plant, total mass, SN, and average SM were measured. The average SM and the SN for each plant was graphed against each other (Fig. 4.2). There was a wide range in both variables with the greater variation found in SN. While the general impression could be that SN was the dominant variable accounting for yield differences among plants, a close look at the results leads to a different conclusion. When the plant population is segregated into subgroups of approximately equal total plant mass, there was a strong negative relationship between SM and SN within each subgroup. That is, there was a high degree of component compensation between SM and SN and yield was not directly dependent solely on either SM or SN.

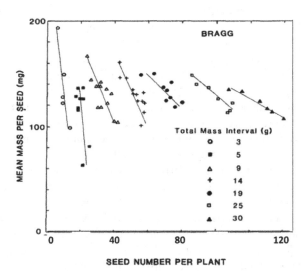

Fig. 4.2 Graph for individual cv. 'Bragg' soybean plants based on their average seed mass and total seed number (Spaeth and Sinclair 1984). The plant population was segregated into subgroups of similar total plant mass revealing a high degree of component compensation of SM and SN within each subgroup as shown by the linear regressions

Another feature in Fig. 4.2 is that the slopes of the subgroups of SM versus SN were less negative as plant mass becomes larger. The less negative slopes for the plant subgroups with larger mass was also associated with greater SN per plant. As a result, the smaller negative slopes of the larger plants was associated with the fact that harvest index, (SN · SM/plant mass), was essentially stable across all plants. Spaeth et al. (1984) found that for the population of plants shown in Fig. 4.2 that the individual plant harvest index was nearly constant at 0.50 across all plants sizes with somewhat greater variation occurring among the smaller plants. Stability in harvest index among individual plants appears to be a fairly common feature of crop plants (Spaeth et al. 1984).

Overall, there is little evidence that SN is an independent variable that determines plant yield. The fact that SN and seed yield are correlated likely reflects the fact that both seed number and yield are dependent on resource availability whether the resource variable is photosynthate or nutrients. That is, the correlation between seed number and yield should not be assumed to indicate causal relationship between the two variables. Crop yield is ultimately dependent on the amount of resource input and not necessarily how it is packaged in terms of seed mass or seed number.

References

Abbate PE, Andrade FH, Culot JP (1995) The effects of radiation and nitrogen on number of grain in wheat. J Agric Sci Camb 124:352–260

Adams MW (1967) Basis of yield component compensation in crop plants with special reference to the field bean. Phaseolus Vulgaris Crop Sci 7:505–510

Demotezs-Mainard S, Jeuffroy M-H, Robin S (1999) Spikce dry matter and nitrogen accumulation before anthesis in wheat as affected by nitrogen fertilizer: relationship to kernels per spike. Field Crops Res 64:249–259

Egli DB (2019) Crop growth rate and the establishment of sink size: a comparison of maize and soybean. J Crop Improv 33:346–362

Garcia RL, Long SP, Wall GW, Osborne CP, Kimball BA, Nie GY, Pinter PJ Jr, Lamorte RL, Wechsung F (1998) Photosynthesis and conductance of spring-wheat leaves: field response to continuous free-air atmospheric CO_2 enrichment. Plant Cell Env 21:659–669

Mishra SP, Mohapatra PK (1987) Soluble carbohydrates and floret fertility in wheat in relation to population density stress. Ann Bot 60:269–277

Reynolds MP, Pellegrineschi A, Skovmand B (2005) Sink-limitation to yield and biomass: a summary of some investigations in spring wheat. J App Biol 146:39–49

Rotundo JL, Borras L, De Bruin J, Petersen P (2012) Physiological strategies for seed number determination in soybean: biomass accumulation, partitioning and seed set efficiency. Field Crops Res 135:58–66

Savin R, Slafer GA (1991) Shading effects on the yield of an Argentinian wheat cultivar. J Agric Sci 116:1–7

Sibony M, Pinthus MJ (1988) Floret initiation and development in spring wheat (Triticum aestivum L.). Ann Bot 61:473–479

Sinclair TR, Jamieson PD (2006) Grain number, wheat yield, and bottling beer: an analysis. Field Crops Res 98:60–67

Sinclair TR, Jamieson PD (2008) Yield and grain number of wheat: a correlation or a causal relationship? Authors' response to the "The importance of grain or kernel number in wheat: a reply to Sinclair and Jamieson" by RA Fischer. Field Crops Res 105:22–26

Slafer GA (2003) Genetic basis of yield as viewed from a crop physiologist's perspective. Ann Appl Biol 142:117–128

Spaeth SC, Sinclair TR (1984) Soybean seed growth. II. Individual seed mass and component compensation. Agron J 76:128–133

Spaeth SC, Randall HC, Sinclair TR, Vendeland JS (1984) Stability of soybean harvest index. Agron J 76:482–486

Yang H, Dong B, Wang Y, Oiao Y, Shi C, Jin L, Liu M (2020) Photosynthetic base of reduced grain yield by shading stress during the early reproductive stage of two wheat cultivars. Sci Rep 10:14353

Chapter 5
Plant Nitrogen Use Efficiency

Nitrogen often limits crop growth and yield as discussed in previous chapters. Due to this limitation, it is not surprising that there is intense interest to maximize crop yield for every unit of applied fertilizer to a field. This interest results from several reasons including the high cost of nitrogen fertilizer in crop production, the need to decrease nitrogen pollution of subsurface and surface water supplies (https://www.epa.gov/nutrientpollutions), and the impact of gaseous oxide release to the atmosphere as a major contributor to global warming (Battye et al. 2017). Therefore, improved crop nitrogen use efficiency (NUE) is seen as a high priority in virtually all crop production.

5.1 Defining Nitrogen Use Efficiency

Research progress in resolving approaches to improved NUE has been problematic with one of the main problems being that NUE is defined in several ways; that is, it is a polysemyous phrase. In fact, Xu et al. (2012) identified at least seven different definitions of NUE that are used in the scientific literature. The various definitions of NUE require a reader to be especially conscientious in being aware of the exact definition and methodology used in every discussion. The various definitions are not equivalent and their relevance in crop improvement can be quite variable.

The simplest definition for NUE for which values can be calculated is the ratio of grain yield (Y) divided by amount of applied fertilizer (N).

$$NUE = Y/N \qquad (5.1)$$

Caution: Phenomenological

This equation could be labelled as a **phenomenological expression**, because NUE is defined by observational data for two input variables. However, neither the denominator nor numerator in Eq. (5.1) offers any access to the mechanisms that may actually impact nitrogen use. The denominator in particular is simply the amount of nitrogen that happened to be applied in a specific field and offers no insight about the fate of the nitrogen once in the soil, uptake by the plants, or its accumulation and distribution in the plants. In spite of the lack of insight about controlling processes that can be gained from Eq. (5.1), it use is prevalent in the crop science literature (Tiong et al. 2021; Fan et al. 2016; Hu et al. 2015).

A more expansive equation for NUE that may help to give more insight is the following:

$$
\begin{aligned}
\text{NUE} =&(\text{Applied N} + \text{Native Soil N}) \cdot \text{Fraction Soil N Available to Plants} \\
&\cdot \text{Fraction Plant N Uptake} \cdot \text{Fraction Plant N in Grain} \\
&/(\text{Applied N} + \text{Native Soil N})
\end{aligned}
\tag{5.2a}
$$

By cancelling the (Applied N + Native Soil N) term in the numerator and denominator, Eq. (5.2a) simplifies to

$$
\begin{aligned}
\text{NUE} =&\text{Fraction Soil N Available to Plants} \\
&* \text{ Fraction Plant N Uptake} \\
&* \text{Fraction Plant N in Grain}
\end{aligned}
\tag{5.2b}
$$

5.2 Nitrogen Use Efficiency Components

Equation (5.2b) offers a description of the three components in nitrogen use to obtain an estimate of NUE. For example, the variables in this expression may have approximate values of Fraction Soil N Available to Plants = 0.4, Fraction Plant N Uptake =

0.9, and Fraction Plant N in Grain $= 0.8$. Consequently, NUE is calculated to equal 0.29, indicating that about 1/3 of the initially available N ends up being harvested in the grain.

In this example (and in the real world), the largest negative impact on NUE is Fraction Soil N Available to Plants. This is a result of the ephemeral nature of inorganic N in the soil. There can be large losses of N to the environment due to its presence in soil runoff, leaching through the soil, and denitrification due to soil microbial activity. The exact amount of the losses from the soil is dependent on environmental conditions and the temporal dynamics of weather events and soil water status. However, under many circumstances the fraction of recovery of soil nitrogen can be even less than 0.4. While it would be very attractive to increase the amount of soil N that is ultimately available to plants, there are currently few practical proposals to substantially decrease soil N losses in field crop production.

On the other hand, plants currently have very high capacities for taking up the nitrogen available to them from the soil as expressed as Fraction Plant N Uptake. Plants have several pathways for N acquisition, including a High-Affinity Transport System (HATS) and a Low-Affinity Transport System (LATS) (Siddiqi et al. 1990). The existence of multiple uptake systems allows plants to be functional in accumulating nitrogen from the soil over the wide range of N concentrations that exist in fertilized soil during a growing season.

Nitrogen uptake pathways are, however, highly regulated as a result of plant activity. Nitrogen accumulation is synchronized with the capacity of the plant to metabolize and utilize nitrogen for growth processes, and also to some extent store nitrogen. The high degree of effectiveness of the regulatory system results, in large part, from communications between nitrogen utilization processes among different destinations in the plant and the uptake systems in the roots. The likely mechanism for communication involves nitrogen cycling within the plant's vascular system. In such a system, illustrated in Fig. 5.1, soluble nitrogen compounds circulate between the roots and shoot through the xylem and phloem (Rufty et al. 1982; Cooper and Clarkson 1989). When circulating nitrogen in the phloem is high, negative feedback on nitrogen uptake in roots occurs. The negative feedback regulatory loop limits nitrate and ammonium uptake in both the HATS and LATS uptake systems (Glass 2003). As a result of negative feedback, inorganic nitrogen acquisition by crop plants remains coordinated with the capacity to utilize N in plant tissue even at times when inorganic soil N in the soil may be high following fertilizer applications.

Despite the abundant evidence indicating strong feedback control over uptake from the soil, there continues to be studies with the goal to improve the uptake system by modifying transport proteins alone (Plett et al. 2018; Noguero and Lacombe 2016). One recent example is a study on the possibility of overexpression of alanine aminotransferase to increase N uptake, but this proved not to increase total nitrate accumulation in several crop species (Tiong et al. 2021). An interesting outcome of this study, however, was that plant vegetative mass was increased. Since many metabolic alterations were found in the transformed plants, it was not possible to identify the exact process that triggered plant mass accumulation.

Fig. 5.1 Illustration of the nitrogen circulation system in the vascular tissue of the plant that results in regulation of both nitrogen uptake associated with the roots and nitrogen metabolism into plant components mainly in the shoot (Rufty and Sinclair 2020)

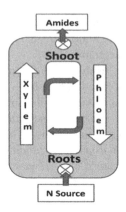

5.3 Plant Nitrogen Accumulation

Given the importance of nitrogen accumulation in achieving high crop yield, one option for increasing N uptake is to select plants that extend the period in the growing season of uptake into the reproductive phase of plant development. Commonly little N is accumulated after anthesis due to a combination of limited N remaining in the soil by the time of anthesis and an apparent decline in root capacity for N assimilation. Evidence has been presented, however, that in selected maize (*Zea mays* L.) hybrids (Ding et al. 2005; Rajcan and Tollenaar 1999) and cultivars of wheat (*Triticum aestivum* L.) (Bogard et al. 2010; Kichey et al. 2007) N accumulation can occur post-anthesis, likely before high rates of seed growth are initiated. To successfully extend nitrogen accumulation into the post-anthesis period, fertilizer management practices are required to provide nitrogen in the soil for plant uptake late in the growing season. Such management practices will probably require methods to provide supplementary nitrogen in the soil at anthesis or minimization of nitrate losses through the use of slow-release fertilizers.

Much of the N accumulated by plants is readily assimilated into N-containing compounds, especially proteins, and sequestered in various plant tissues. To substantially increase plant N uptake it is necessary to alter plants to increase the amount of N-storage tissues or the N-storage capacity of tissues, separate from involvement in the circulating regulatory system. Leaves would be especially important in N storage due to the large amounts of proteins in the leaves, particularly Rubisco, which is quite stable and can be readily catabolized for transfer of N to elsewhere in the plant.

The capacity for storage of N in crop plants seems, however, to be limited. This limitation is illustrated by the relationship observed between crop nitrogen accumulation and crop mass (Fig. 5.2). The relationship differs somewhat among species due to differences in nitrogen concentration of the various plant tissues of each species. The curves initially increase rapidly due to the high proportion of nitrogen-rich leaves of the total plant mass. As the stem mass becomes an increasing fraction of the total plant mass, the increase in plant nitrogen is at a lower rate due to the lower nitrogen

Fig. 5.2 Graphs of crop
nitrogen uptake versus crop
mass based on measurements
for pea (*Pisum sativum* L.),
wheat (*Triticum aestivum*
L.), and maize (Zea mays L.)
reported by Lemaire et al.
(2007)

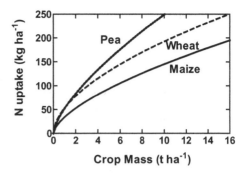

concentration of the stems. Therefore, the resulting relationship between nitrogen
accumulation and crop mass during crop growth is curvilinear with decreasing rate
of nitrogen accumulation as crop mass increases.

The final component in defining NUE in Eq. (5.2b) is the Fraction Plant N in Grain,
which is equivalent to the Nitrogen Harvest Index (NHI). During seed growth only
small amounts of new nitrogen is usually accumulated by a crop. Therefore, to provide
N for the growing seeds, N stored in vegetative tissues, mainly as protein, must
generally be catabolized during grain fill and the released amino acids transported
to the growing grains (Kant 2018) as illustrated in Chapter 3 (Fig. 3.2). The result of
the protein catabolism in the leaves is that plant photosynthetic capacity decreases
due to decreasing leaf area resulting from leaf senescence and/or to decreasing leaf
nitrogen concentration. This process of "self destruction" (Sinclair and de Wit 1975)
defines the shift in nitrogen from vegetative tissue to growing seeds resulting in a
loss in physiological activity in the vegetative tissues.

There have been studies with transgenic plants evaluating and attempting to
improve transport of nitrogen to developing sink tissues (Chen et al. 2020; Perchlik
and Tegeder 2017). However, these studies failed to recognize the critical variable of
plant NHI in assessing potential benefits in increasing seed production. Without an
increase in NHI, it is difficult to conclude that increased transfer rate of N to developing seeds is the basis of a seed yield increase. In fact, with modern commercial
varieties and management practices, N remobilization from vegetative tissue to seeds
is very high and NHI is typically 0.80 or more (Bogard et al. 2010; Kichey et al.
2007). Since some N is required in structural proteins of vegetative tissue, options
for increasing NHI to values much greater than 0.80 seem extremely limited.

Overall, the two plant variables for altering NUE—Fraction Plant N Uptake and
Fraction Plant N in Grain—seem to offer little opportunity for improvement. Uptake
is restrained by feedback control due to coordination with whole plant growth rates
and NHI is already very high in commercial crop cultivars as a consequence of
intensive plant breeding and selection. On the other hand, the remaining component
of NUE, which is Fraction Soil N Available to Plants, is relatively small and it has
thus far been difficult to achieve meaningful increases much greater than about 0.4.
Improved fertilizer technology and soil management appear to be the most fruitful
paths for increasing overall crop NUE.

References

Battye W, Aneja VP, Schlesinger WH (2017) Is nitrogen the next carbon? Earth's Future 5:1–11

Bogard M, Allard V, Rancourt-Hulmel M, Heumez E, Mache J-M, Jeuffroy M-H (2010) Deviation from grain protein concentration-grain yield negative relationship is highly correlated to post-anthesis N uptake in winter wheat. J Exp Bot 61:4303–4312

Chen KE, Chen HY, T CS, Tsay YF (2020) Improving nitrogen use efficiency by manipulating nitrate remobilization in plants. Nat Plants 6:1126–1135

Cooper HD, Clarkson DT (1989) Cycling of amino-nitrogen and other nutrients between shoots and roots in cereals—a possible mechanism integrating shoot and root in the regulation of nutrient uptake. J Exp Bot 40:753–762

Ding L, Wang KJ, Jiang GM, Biswas DK, Xu H, Li LF, Li YH (2005) Effects of nitrogen deficiency on photosynthetic traits of maize hybrids released in different years. Ann Bot 96:925–930

Fan X, Tang Z, Tan Y, Zhang Y, Luo B, Yang M, Lian X, Shen Q, Miller AJ, Xu G (2016) Overexpression of a pH-sensitive nitrate transporter in rice increases crop yields. Proc Nat Acad Sci 113:7118–7123

Glass ADM (2003) Nitrogen use efficiency: physiological restraints upon nitrogen absorption. Crit Rev Plant Sci 22:453–470

Hu B, Wang W, Ou S, Tang J, Li H, Che R et al (2015) Variation in NRT1.1B contributes to nitrate-use divergence between rice subspecies. Nat Gen 47:834–838

Kant S (2018) Understanding nitrate uptake, signaling and remobilization for improving plant nitrogen use efficiency. Seminars Cell Dev Biol. 74:89–96

Kichey T, Hirel B, Heumez E, Dubois F, Le Gouis J (2007) In winter wheat (*Triticum aestivum* L.), post-anthesis nitrogen uptake and remobilization to the grain correlates with agronomic traits and nitrogen physiological markers. Field Crops Res 102:22–32

Lemaire G, van Oosterom E, Sheehy J, Jeuffroy MH, Massignam A, Rossario L (2007) Is crop N demand more closely related to dry matter accumulation or leaf area expansion during vegetative growth. Field Crops Res 100:91–106

Noguero M, Lacombe B (2016) Transporters involved in root nitrate uptake and sensing by Arabidopsis. Front Plant Sci 7:1391

Perchlik M, Tegeder M (2017) Improving plant nitrogen use efficiency through alteration of amino acid of amino acid transport processes. Plant Physiol 175:235–247

Plett DC, Holtham LR, Okamoto M, Garnett TP (2018) Nitrate uptake and its regulation in relation to improving nitrogen use efficiency in cereals. Sem Cell Devel Biol 74:97–104

Rajcan I, Tollenaar M (1999) Source: sink ratio and leaf senescence in maize: II. Nitrogen metabolism during grain filling. Field Crops Res 60:255–265

Rufty TW, Volk RJ, McClure PR, Israel DW, Raper CD Jr (1982) The relative content of NO_3^- and reduced-N in xylem exudate as an indicator of root reduction of concurrently absorbed $^{15}NO_3^-$. Plant Physio 69:166–170

Rufty TW, Sinclair TR (2020) Cycling of amino nitrogen and other nutrients between shoots and roots in cereals—a possible mechanism integrating shoot and root in the regulation of nutrient uptake by Cooper HD, Clarkson DT. J Exp Botany (1989) 40:753–762; Crop Sci 60:2192–2194

Sinclair TR, de Wit CT (1975) Photosynthate and nitrogen requirements for seed production by various crops. Science 190:565–567

Siddiqi MY, Glass ADM, Ruth TJ, Rufty TW Jr (1990) Studies of the uptake of nitrate in barley I. Kinetics of $^{13}NO_3^-$ influx. Plant Physio 93:1426–1432

Tiong J, Sharma N, Sampath R, Mackenzie N, Watanabe S, Metol C, Lu Z, Skinner W, Lu Y, Kridl J, Baumann U, Heuer S, Kaiser B, Okamoto M (2021) Improving nitrogen use efficiency through overexpression o alanine aminotransferase in rice, wheat, and barley. Front Plant Sci 12:628521

Xu G, Fan X, Miller AJ (2012) Plant nitrogen assimilation and use efficiency. Ann Rev Plant Biol 63:153–182

Chapter 6
Osmolyte Accumulation

One of the major challenges in crop production is minimization of the impact of soil water deficit on crop development, growth, and yield. Osmolyte accumulation in plant cells has been one approach often suggested for improving crop performance during water-deficit conditions. In fact, investigations on possible benefits of osmolyte accumulation in crop plants has been reported for at least 90 years. An early glimpse of results from succeeding studies was presented by Martin (1930), who compared measurements of osmotic potential in sorghum and maize. He concluded that differences in osmolyte accumulation in the leaves of the two species "could not account for any superior drought resistance of sorghums".

Numerous reviews have examined the results of studies on osmolyte accumulation in crop species and reached mixed conclusions about possible benefits. In the review of Serraj and Sinclair (2002), it was concluded that most studies had not found yield benefit from osmotic adjustment although there were cases of positive responses. The review by Blum (2017) identified 24 of 26 selected studies in which a positive response to osmotic adjustment was reported. However, ambiguity exists with this conclusion because the basis for selection of the 26 studies was not explained. The bottom line as concluded by Turner (2018) was that with greater than 500 publications on osmolyte accumulation in various crop species "there is only one example ... in which a high-OA [osmolyte accumulation] cultivar (of bread wheat) has been successfully bred for water-limited environments." In this single case, the yield increase at the driest sites was 10%.

Given the ambiguity about the possible benefit of osmolyte accumulation, skepticism is obviously called for when considering expectations for crop improvement under water-deficit conditions. Of course, if an approach to obtain consistent positive response to osmolyte accumulation is identified, then this could provide another tool in the development of drought-resilient plants. In this review, the possible roles of osmolyte accumulation in increasing crop yield under water deficit are considered.

6.1   Soil Water-Deficit Stress Under Field Conditions

Many studies of the impact of osmolyte accumulation on plant performance have been done under highly controlled and regulated conditions. One popular approach in attempting to mimic water deficit is to expose plants or specific plant tissue to external osmotic stress. That is, roots or specific tissues are immersed in a solution containing an osmoticum to impose a decreased water potential. The problem with this approach is that this is essentially a 'shock treatment' with no opportunity for plant acclimation and adjustment in response to the stress. Such a shock treatment is unlike any stress in the field where soil water deficits develop over prolonged periods of time. Furthermore, experimental exposure to osmolytes in solution may result in osmolyte penetration of membranes, especially with extended exposures, which can further disrupt the normal physiological progression or plant acclimation to soil water deficit. While osmotic-stress experiments might give some physiological perspective in some circumstances, considerable caution is required in extrapolating the results from such treatments to plant response to progressive soil drying.

Experimental approaches involving drying of solid rooting media also need to be evaluated with caution. An immediate issue may result from the type of rooting media used in an experiment. The water release of artificial or constructed rooting media are generally unlike that of mineral soil. Hence, the development of water deficit on these non-mineral media are likely to be unlike that experienced by plants grown in the field. For instance, artificial media high in sand fraction (>60%) results in very rapid decreases in plant transpiration rate once the media dries to the threshold of partial stomata closure (Sinclair et al. 1998). On the other hand, artificial media containing high fractions of organic matter express thresholds for partial stomatal closure very early in the rooting media drying cycle (Wahbi and Sinclair 2007).

Caution: Time

Another concern in experiments using solid rooting media, usually in pots, is the **temporal dynamics** in the establishment of water-deficit treatments. To simplify water-deficit research, many studies employ a protocol to rapidly dry the rooting media to a relatively stable water-deficit treatment and then the water-deficit treatment is held at a particular soil water content by controlled watering. This approach is used in pot experiments done in controlled environments and in the field with controlled irrigation regimes. While these 'static' treatments may offer some physiological insight about plant behavior, the results do not necessarily relate to plant response to the day-to-day changes in soil water content in the field. Crops in the field are subjected to continuous changes in soil water resulting from both abrupt water additions due to precipitation and irrigation, and from relatively slow water removal by transpiration, soil evaporation, and soil water percolation. A key factor contributing to the large variation in results from osmolyte experiments could be the differences among experiments in the temporal patterns in soil water deficits.

Rather than the imposition of 'static' water deficit treatments, an alternate approach is to track plant response as the soil is allowed to dry progressively. In our experience of studying plant response to soil drying, it is necessary to allow water deficit to develop slowly over an extended time period. We allow water deficits to develop over at least 8–10 days, which results in repeatable plant-drought responses (Chiango et al. 2022; Manandhar et al. 2017; Gholipoor et al. 2013). More rapid drying, which is not uncommon in many studies, seems to result in some type of 'shock' response to the imposition of water deficit.

The slow-drying approach in pots allows the soil to dry from 'pot capacity' (analogous to field capacity) to a stress endpoint at which stomatal conductance reaches a minimum value. The pot is weighed at least daily during the dry-down to determine soil water status. Transpiration rate (achieved by preventing soil evaporation) is calculated as the difference in weight between successive weighing. To account for differences in plant size and in the daily environment, transpiration rates of the drying pots on each day are normalized (NTR) against average transpiration rates of well-watered pots. The fraction of transpirable soil water (FTSW) existing in the pot at any time during the dry-down is calculated based on the ratio of the current soil water level relative to the total transpirable water that can be stored in the soil. The total transpirable soil water is calculated as the weight difference between the initial weight and the endpoint weight when transpirable soil water is depleted. In our studies with annual crop species, the endpoint occurs when the decrease in NTR is no longer linearly related to decreasing FTSW, which commonly occurs at NTR$\approx$0.1.

A typical dry-down pattern obtained using the NTR and FTSW approach is shown in Fig. 6.1. In this graph, soil drying is depicted going from left to right. As is commonly found, soil drying can be described statistically by two linear segments that intersect at a breakpoint (BP), with the breakpoint indicative of the FTSW at the initiation of stomata closure. In this case, the BP for this maize hybrid was FTSW = 0.37, which is in the range of BP values of 0.25 to 0.40 commonly observed for transpiration response in a number of experimental studies with crop species (Sadras and Milroy 1996).

Fig. 6.1 Graph of
normalized transpiration rate
(NTR) versus fraction
transpirable soil water
(FTSW) during a dry-down
experiment for a maize
hybrid (Gholipoor et al.
2013)

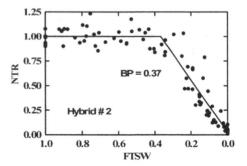

Once soil dries to the severe stress level of NTR < 0.1 in Fig. 6.1, stomata conductance is at its minimum value, CO_2 exchange and photosynthesis are essentially fully restricted, and the plant enters a 'survival' mode. While considerable research has been done under such severe water deficits representative of the survival mode, this condition is generally not relevant for crop production. In the field, it takes considerable time to reach the survival stage during which the loss in crop production is continually accumulating. Even if the crop survives the severe drought, the accumulating structural loss (leaf senescence) and decreasing biochemical activity will severely restrain yields. The very large yield loss when a crop experiences survival drought, even if it survives, will in almost all cases be economically devastating for growers. Plant survival traits are not generally relevant for annual crop production.

6.2   Hypotheses for Benefit of Osmolyte Accumulation

In the plant literature, three overarching hypotheses have been offered as possibilities for how osmolyte accumulation might improve crop performance under water-deficit conditions. These hypotheses are (1) osmolyte protection of cell components, (2) osmolyte maintenance of plant tissue turgor under soil water-deficit conditions, and (3) osmolyte enhancement of soil water extraction.

(1)   Protection of cellular components

Accumulated osmolytes have been proposed to interact with cell components to prevent their denaturation or destruction as cells desiccate. Such protection resulting from the accumulation of various osmolytes has been summarized by Suprasanna et al. (2016). However, like a survival trait, the loss of cell turgor that might cause drought damage occurs only very late in the soil drying cycle. For example, in a field study with maize hybrids, an osmotic-adjusting, near-isoline did not show leaf osmolyte accumulation until the severe stress level of $\psi \leq -1.2$ MPa was reached (Chimenti et al. 2006). As illustrated in Fig. 6.2 for three soybean genotypes, osmotic potential does not usually decline until the plant is nearing the survival stage of FTSW ≤ 0.1. This delayed response pattern in bulk leaf water status is likely a consequence

Fig. 6.2 Graph of leaf osmotic potential (OP) of three soybean genotypes versus fraction of transpirable soil water (FTSW) during soil drying (Bagherzadi et al. 2017)

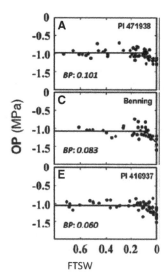

of the partial stomata closure that occurs earlier in the soil water-deficit cycle (i.e. higher FTSW). The stomata closure prevents desiccation of the bulk leaf so that it is actually maintained in a 'well-watered' state during much of the period of developing soil water deficit. Hence, a benefit of cellular component protection from osmolyte accumulation could likely occur only when the plant is approaching the survival stage and yield capacity has already been greatly diminished.

(2) Maintenance of cell turgor

The proposed mechanism for maintenance of cell turgor is osmolyte accumulation in cells causing water to move across membranes into cells. The greater turgor as a result of osmolyte when soil water deficit develops is suggested to allow sustained physiological activity of the cell, and by extension, a benefit for the entire tissue. At the leaf level, the expectation is that turgor maintenance will result in avoidance or delay in the onset of stomata closure and leaf wilting.

One of the difficulties in evaluating the cell turgor hypothesis is that turgor effects occur at the cell level, and turgor measured for the bulk tissue may not reflect the status of individual cells. For example, the mechanics of stomata conductance depend specifically on changes in the turgor of the guard cells and of the neighboring epidermal cells, not bulk leaf turgor. As discussed previously, initiation of stomata closure at relatively high soil water content causes bulk leaf water status to be maintained until severe soil water deficit is reached. In a study with sorghum in which leaf rolling is closely linked to bulliform cell turgor, bulk leaf water potential was decreased substantially to ≤ -1.0 MPa during diurnal observations before leaf rolling was observed (Wright et al. 1983). In this instance and in others, there is very little information on osmolyte accumulation in the specific cells of relevance.

More specifically, a mechanism that maintains guard cell turgor to sustain stomata opening may well be exactly the wrong response that would be advantageous under many circumstances for crops during the development of progressive soil water deficit. Continued aggressive extraction of soil water with open stomata as the soil dries would result in continued, rapid decline in soil water content that would eventually place the crop in more peril. In contrast, a prudent response for a plant confronted with drying soil may be to initiate partial stomata closure and soil water conservation early in soil drying so that physiological activity, even at a lower rate, can be prolonged if drought conditions persist. The relative merit of partial stomata closure early in soil drying was indicated in simulations of soybean production across the US (Sinclair et al. 2010). It was found that there was a high probability of yield increase in nearly all regions of the US with early stomata closure with soil drying as compared with delayed stomata closure proposed as a benefit from osmolyte accumulation (Fig. 6.3).

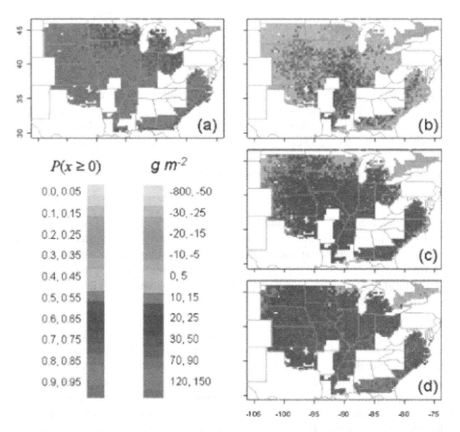

Fig. 6.3 Results of simulations comparing soybean yield increase resulting from early stomata closure with soil drying as compared to delayed stomata closure (Sinclair et al. 2010). Map **a** is the probability (P) of yield increase with early closure relative to delayed stomata closure. Absolute yield increase (g m^{-2}) is shown in maps for the **b** 75% percentile ranking (wet), **c** 50% percentile ranking (median), and **d** 25% percentile ranking (dry)

A possibility for an advantage resulting from turgor maintenance as a result of osmolyte accumulation may occur in root tips. In this case, greater turgor in the root tips could support greater extension rates of roots as observed in studies with young seedlings (Greacen and Oh 1972; Sharp and Davies 1979). If there is water in deeper soil layers, then greater rates of root extension could allow greater access to the deeper water. However, there have been very few studies on root-tip osmotic pressure and increased root extension in more mature plants. In one study (Beseli et al. 2019), root growth was examined during vegetative development with two maize, near-isoline hybrids differing is leaf osmolyte accumulation. They found, however, no consistent difference between the near-isolines in their root-tip osmotic pressure or in root length extension at any time during plant development.

Also, the hypothesis that increased root extension rate is an advantageous trait may not have widespread application. The uncertainty of the impact of more rapid root extension rate can result from the temporal pattern of water use. Increased root extension rate could increase access to water in deeper soil layers during the early-season vegetative stage that would allow enhanced plant growth. The difficulty, however, is that these larger plants would have consumed a greater fraction of the seasonal soil water during the vegetative stage leaving the crop more vulnerable to late-season water deficit. As a result of the increased root extension rate and early-season consumption of water, the crop would be more vulnerable to late-season water deficits when seed growth is occurring. The overall impact on yield is not clear cut. In the simulation study of soybean across the US, Sinclair et al. (2010) compared yields for two rates of root depth extension (20 and 25 mm d^{-1}). Their results gave a probability of yield decrease for the more rapid root extension rate as compared to the slower extension rate in the US except in the driest regions.

(3) Enhanced soil water extraction

It has been hypothesized that osmolyte accumulation in leaves could allow for an increased amount of water uptake by increasing the hydraulic potential gradient between the plant and the soil. However, the amount of water available for plant extraction beyond that which could be originally extracted by the plant with less osmolyte accumulation is very small in most soils. That is, soil water-release curves show that little water is available for increased water extraction at soil water potential less than something in the range of -1.5 MPa.

Another hypothesis for osmolyte accumulation to enhance soil water extraction relates to the rate of extraction. An increased hydrostatic potential gradient between the plant and soil would allow for a faster rate of water uptake by the crop when the soil is 'wet'. However, the hypothesized advantage of the faster water uptake rate needs to be assessed within a temporal context. More rapid water uptake when the soil is 'wet' also means that the soil will dry faster. That is, the advantage of faster water uptake by osmolyte accumulation also has the risk of the soil drying more quickly to levels that negatively impact plant activity. In particular, more rapid soil water extraction by osmolyte accumulation early in the growing season may make the crop more vulnerable to drought later in the season during flowering and pod fill, which are crucial in determining crop yield.

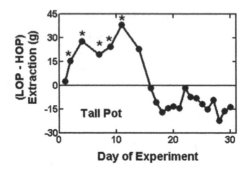

Fig. 6.4 Differences in daily soil water extraction in an 8.4 L pot experiment between an osmolyte accumulating near-isoline (LOP, low osmotic potential) and a non-osmolyte-accumulating near-isoline (HOP, high osmotic potential) graphed against days of the experiment (Beseli et al. 2020). Values of LOP-HOP greater than 0 on the abscissa in the first two weeks of the experiment reflect greater water use rates by the osmolyte accumulating near-isoline. Asterisks indicate significant differences on that day as judged by a P-value ≤ 0.05

The impact on temporal dynamics of soil water extraction as a result of osmolyte accumulation was illustrated in a study of two near-isoline, maize hybrids that differed in leaf osmotic potential (Beseli et al. 2020). The difference in transpiration rate throughout a side-by-side soil dry-down in a pot experiment is shown in Fig. 6.4. Initially, transpiration rate was greater in the osmolyte-accumulating hybrid than the non-osmolyte-accumulating hybrid as hypothesized. However, as the soil dried the non-osmolyte-accumulating near-isoline with sustained transpiration rate eventually had the greater transpiration rate. In this particular experiment in which pot volume was 8.4 L, the non-osymolyte-accumulating hybrid actually had the favorable response of reaching NTR ≤ 0.1 three days later than the osmolyte-accumulating hybrid. In an experiment with 11 L pots and a more humid environment, the endpoint of NTR ≤ 0.1 was reached by the non-osymolyte-accumulating hybrid six days later than the osymolyte-accumulating hybrid. Any relative advantage of the osmolyte-accumulating characteristic depended on the temporal pattern of soil water availability and the extent and duration of plant exposure to deficit soil water.

6.3   A Skeptics Conclusion

Osmolyte accumulation has been the focus of study and speculation for nearly a century. Yet, the impact of the accumulation of crop yield remains unresolved. This ambiguity seems in large part a result of a failure to consider the transient nature of soil water availability through the growing season. Soil water content is changing as water is added and removed from the soil. There is no static condition on which to judge the impact of osmolyte accumulation. Much attention has been given to plant performance under severe drought often associated with plant survival. In agricultural production, plant survival of water deficit is essentially irrelevant since for the grower crop survival does little to ease the economic catastrophe. Altered timing in crop water extraction from the soil during the growing season seems particularly important to increase yield if soil water deficit develops. Diminished water use early in the growing season in favor of water availability during reproductive stages offers a high probability of yield increase. In fact, it appears that many suggestions for osmolyte accumulation are likely to result in negative or no impact on crop yield.

References

Bagherzadi L, Sinclair TR, Zwieniecki M, Secchi F, Hoffman W, Carter TE, Rufty TW (2017) Assessing water-related plant traits to explain slow-wilting in soybean PI 471938. J Crop Improv 31:400–417

Beseli A, Hall AJ, Manandhar A, Sinclair TR (2019) Root osmotic potential and length for two maize lines differing leaf osmotic potential. J Crop Improv 33:429–444

Beseli A, Shekoofa A, Ali M, Sinclair TR (2020) Temporal water use by two maize lines differing in leaf osmotic potential. Crop Sci 60:945–953

Blum A (2017) Osmotic adjustment is a prime drought stress adaptive engine in support of plant production. Plant Cell Environ 40:4–10

Chiango H, Jafarikouhini PD, Figueiredo A, Silva J, Sinclair TR, Holland J (2022) Drought resilience in CIMMYT maize lines adapted to Africa resulting from transpiration sensitivity to vapor pressure deficit and soil drying. J Crop Improv 36:301–315

Chimenti CA, Marcantonio M, Hall AJ (2006) Divergent selection for osmotic adjustment results in improved drought tolerance in maize (Zea mays L.) in both early growth and flowering phases. Field Crops Res 95:305–315

Greacen EL, Oh JS (1972) The physics of root growth. Nat New Biol 235:24–25

Gholipoor M, Sinclair TR, Raza MAS, Loffler C, Cooper M, Messina CD (2013) Maize hybrid variability for transpiration decrease with progressive soil drying. J Agron Crop Sci 199:23–29

Manandhar A, Sinclair TR, Rufty TW, Ghanem ME (2017) Leaf expandion and transpiration response to soi ldrying and recovery among cowpea genotypes. Crop Sci 57:2109–2116

Martin JH (1930) The comparative drought resistance of sorghums and corn. Agron J 22:993–1003

Sadras VO, Milroy SP (1996) Soil-water thresholds for the responses of leaf expansion and gas exchange: a review. Field Crops Res 47:253–266

Serraj R, Sinclair TR (2002) Osmolyte accumulation: can it really help increase crop yield under drought conditions? Plant Cell Environ 25:333–341

Sharp RE, Davies WJ (1979) Solute regulation and growth by roots and shoots of water-stressed maize plants. Planta 147:43–49

Sinclair TR, Hammond LC, Harrison J (1998) Extractable soil water and transpiration rate of soybean on sandy soil. Agron J 90:363–368

Sinclair TR, Messina CD, Beatty A, Samples M (2010) Assessment across the United States of the benefits of altered soybean drought traits. Agron J 102:475–482

Suprasanna P, Nikalje GC, Rai AN (2016) Osmolyte accumulation and implications in plant abiotic stress tolerance. In: Iqbal et al. (eds) Osmolytes and plants acclimation to changing environment: emerging omics technologies. Springer, India

Turner NC (2018) Turgor maintenance by osmotic adjustment: 40 years of progress. J Exp Bot 69:3223–3233

Wahbi A, Sinclair TR (2007) Transpiration response of arabidopsis, maize, and soybean to drying of artificial and mineral soil. Environ Exp Bot 59:188–192

Wright GC, Smith RCG, Morgan JM (1983) Differences between two grain sorghum genotypes to adaptation to drought stress. III. Pysiological responses. Aust J Agric Res 34:637–651

Chapter 7
Plant Water Use Efficiency

Soil water deficits very often develop in crop growing seasons, depressing plant development and growth, and ultimately limiting final crop yield. A popular slogan for research to overcome these water limitations is 'more crop per drop'. While this slogan for improved water use efficiency is catchy and seems a reasonable goal, are there actually realistic approaches for plant improvement to give more crop per drop?

Caution: Phenomenological

7.1 Phenomenological View

One framework suggesting insight for improving yield under limiting water is the following phenomenological equation presented by Passioura (1977).

$$Y = T \cdot HI \cdot WUE \tag{7.1}$$

T. Sinclair and T. W. Rufty, *Bringing Skepticism to Crop Science*, SpringerBriefs in Agriculture, https://doi.org/10.1007/978-3-031-14414-1_7

where yield (Y) is described as a function of the amount of soil water extracted for transpiration (T), plant harvest index (HI), and water use efficiency (WUE). Of course, the value of T can be no larger than the amount of water input to the soil. Further, HI in existing modern crop species appears to have reached near-maximum values. Significantly, HI values are fairly stable over a wide range of environmental conditions except in cases of severe drought stress (Spaeth et al. 1984; Sinclair et al. 1990). Consequently, the WUE term remains as the apparent key variable in Eq. (7.1) to achieve increased yield.

Much research to improve WUE has been focused at the leaf level and based on the CO_2 concentration gradient between the atmosphere and the interior of the leaf (Rizza et al. 2012; Vadez et al. 2014). The logic of this approach is that large gradients of CO_2 between the atmosphere and the leaf interior will reflect greater leaf photosynthesis capacity, and hence, the ratio of photosynthesis rate to transpiration rate is increased. Two plant behaviors have been identified to achieve the overall desired increase in CO_2 gradient as a result of decreased leaf interior CO_2 concentration. One behavior is partial stomata closure to diminish the rate of replenishment of CO_2 in the leaf interior so that leaf photosynthesis draws down the CO_2 concentration. Of course, a negative aspect of this behavior is that a decreased leaf interior CO_2 concentration will suppress leaf photosynthesis rate.

A second photosynthetic behavior to increase the CO_2 gradient into the leaf interior is to identify plants that have increased capacity for leaf carbon assimilation. That is, plants are to be found that have especially high leaf photosynthesis capacity to draw down the leaf interior CO_2 concentration. However, as discussed in Chap. 3, CO_2 assimilation rates of current commercial crop cultivars are already high. Unless current cultivars of a species are shown to have low photosynthetic capacity and low yields, there appear to be limited options for increasing photosynthesis activity that actually lower leaf interior CO_2 concentration.

Even though the potential for lowering interior CO_2 concentration seems limited, much research has been done with this goal. The experimental approach that has often been used has focused on measurements of carbon isotope discrimination. This approach is based on differential isotope gas diffusion and discrimination in carboxylation of ^{12}C and ^{13}C (Farquhar et al. 1982). The C isotope discrimination values derived from assimilation equations relate to the ratio of leaf intercellular CO_2 concentration and atmospheric CO_2 concentration. In comparisons of genotypes, an implicit assumption is that the CO_2 concentration ratio between interior and exterior CO_2 concentration is nearly stable for each genotype, or that changes in the ratio through the growing season are in synchrony among genotypes. Two difficulties may exist in these assumptions. One difficulty is that the growth pattern among different tested genotypes may well differ so that leaf gas exchange may not vary in synchrony among genotypes through a growing season and under changing environmental conditions. A second, and possibly more important difficulty, is that the genotypes may differ in sensitivity to vapor pressure deficit so that the temporal pattern in gas exchange will be sensitive to changes in vapor pressure deficit through the course of a day and to day-to-day changes in weather conditions. Not surprisingly, field experiments have generally only been successful in identifying consistent differences among genotypes in carbon discrimination under fairly stable environmental conditions. Stable conditions are typically associated with more arid conditions and less so in more variable humid climate conditions where grain crop species are commonly grown (Sinclair 2012).

7.2 Mechanistic Description of WUE

The importance of atmospheric humidity in analyzing the relationship between plant growth and water loss was shown many years ago by de Wit (1958). He presented an empirical analysis of the considerable amount of data collected worldwide in which plant growth and water loss were measured under a wide range of conditions. These data included a range of environmental conditions, genotypes, and soil water status and fertility. Importantly, a key feature of de Wit's analysis was normalization of plant water loss by dividing plant water loss by the water loss from an open water surface measured in each experiment. A graph of plant growth versus normalized transpiration that included all data collected for each species was highly linear within each species. For the three species shown in Fig. 7.1, the slopes of the graphs differed among species with sorghum (*Sorghum bicolor* L.) having the highest slope, wheat (*Triticum aestivum* L.) an intermediate slope, and alfalfa (*Medicago sativa* L.) the lowest slope. The stability in the slope for each species is directly related to the water use efficiency of each species.

To gain more insight about the results of the analysis presented by de Wit (1958), Tanner and Sinclair (1983) offered a mechanistic derivation of the relationship between grain yield and transpiration water loss for crop canopies starting with equations describing instantaneous leaf gas exchange of CO_2 and water vapor. The starting equations defined at the leaf level represented leaf gas exchange based on the molecular gradients of CO_2 and water vapor and the resistances to their diffusion. Crop mass accumulation was up-scaled from leaf gas exchange using some of the same assumptions about canopy gas exchange processes as presented earlier in Chap. 3 in the derivation of radiation use efficiency. That is, (1) leaf segments are in either direct solar radiation or shaded conditions; (2) there is a horizontally random distribution of leaves; (3) the leaf area exposed to direct radiation (LAI_D)is approximately 1.4; (4) growth respiration rate is defined by the biochemical composition (i.e. carbohydrate, protein, and lipid) of the synthesized plant products of each crop species (Penning de Vries 1975); and (5) maintenance respiration rate is assumed to be approximately equal to the total carbon assimilation of leaves in the shade.

Tanner and Sinclair (1983) argued that for the canopy as a whole, transpiration rate can be calculated based on leaves exposed either to direct solar radiation or to shade solar radiation. To represent their assumption, they approximated the shaded-leaf transpiration rate as equivalent to the transpiration rate of a small leaf area exposed to direct radiation. Their proposed leaf area index for this approximation for estimating shade transpiration rate was 0.8, which was obtained for a canopy

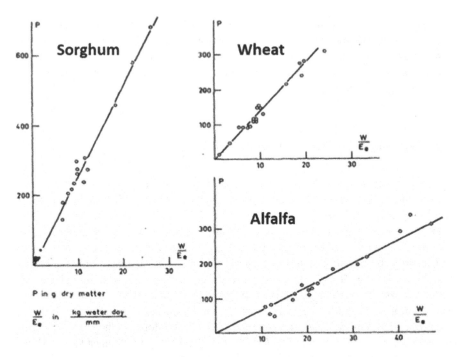

Fig. 7.1 Graphs presented by de Wit (1958) on experimental observations of plant-growth mass labeled as P on the ordinate versus water loss rate divided by open-surface water evaporation labeled as W/E_0 on the abscissa. The data for each species were obtained from published experiments for sorghum, wheat, and alfalfa. The data were obtained for plants grown in containers under a range of differing conditions including several genotypes within each species, soil conditions including fertility and water status, and the physical environment

with a LAI > 3.0 based on the argument that the gas diffusion ratio between leaves exposed to diffuse and to direct radiation is roughly 0.3. Consequently, a suitable effective transpiring leaf area index (LAI_{trans}) based on transpiration rate calculated from direct solar radiation assumed a transpirational leaf area index of 2.2 (1.4 + 0.8).

The result of the above derivation was represented as a time-integrated gas exchange function in a format somewhat similar to Eq. (7.1). However, in the following Eq. (7.2) the fully defined key variables are VPD and k. The variable k is defined by specific, mechanistic properties related to leaf canopy gas exchange.

$$Y = HI \cdot k \cdot \int (T/VPD)dt / \int dt, \qquad (7.2)$$

where

$$k = (abc/1.5)(PC_a/\varrho c)LAI_D/LAI_{trans}$$

where a = molecular ratio of CH_2O/CO_2 (0.68)

b = hexose conversion to plant mass based on plant carbohydrate, protein, and lipid

c = $(1-C_i/C_a)$, where C_i = leaf interior CO_2 concentration, C_a = atmospheric CO_2 concentration (~ 0.3 for C3 species and ~0.7 for C4 species)

P = atmospheric pressure

ϱ = air density

ε = ratio mole weight of water vapor to air

The ratio of several parameters in the definition of k are fairly stable with much of the variation in k among species resulting from the two variables b and c. The value of c is generally fairly constant within each high-yielding crop species, with the major notable difference existing between C3 and C4 species (Wong et al. 1979). The value of b is dependent on the energy content of the synthesized plant mass. The value of b is high for synthesis of plant products high in carbohydrate content, and low for synthesized plant products high in protein and lipid (Fig. 3.1).

Based on the definition of k, an essentially unique value of k is defined for each crop species. For example, the constant value of k is approximately 9 Pa for C4 species such as maize (*Zea mays* L.), 6 Pa for C3 species such as potato (*Solanum tuberosum* L.), and 5 Pa for C3 legume species such as alfalfa (*Medicago sativa* L.) (Tanner and Sinclair 1983). These differences in k are consistent with the slope differences among crop species found by de Wit (1958) (Fig. 7.1).

7.3 Atmospheric Humidity

Caution: Time

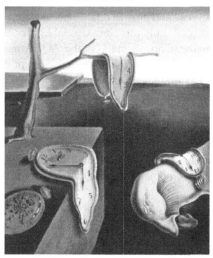

Critically, Eq. (7.2) includes VPD explicitly within an integral as a major variable influencing yield. Given the potential variation in VPD with changing environmental conditions, yield dependence on available water can be quite variable depending on the value of VPD. In particular, large values of VPD can have a major negative impact on yield when soil water is limited. One approach to limit the possible negative impact of large values of VPD is to have plants that minimize transpiration rates under elevated VPD, i.e., partial stomata closure when atmospheric vapor pressure deficit is high. Alternatively, a management solution is to avoid high VPD conditions, which can be done by selecting crop species that can be grown during cool periods of the year when VPD is low.

Recently, a genetic solution has been to develop crop cultivars to express specifically partial stomata closure under elevated VPD. Elevated VPD is likely to occur during the midday resulting in 'midday stomata closure'. For example, in one day VPD can vary from near 0 kPa early in the morning to values that may reach 3 or 4 kPa, or higher in the afternoon. Over longer time periods, weather conditions can vary resulting in potentially large changes in VPD, and hence variable crop transpiration rate. The consequence of partial stomata closure during periods of elevated VPD would be a decrease in the effective VPD for transpiration rate when integrated over an entire day. The important consequence of VDP sensitivity would be conserved soil water, which would be available to sustain plant physiological activity during periods of late-season water deficit.

Genotypic expression of partial stomata closure at elevated VPD, illustrated in the top panel of Fig. 7.2 in a comparison of two maize lines, has been identified in at least a few genotypes of every crop species examined thus far, including soybean, peanut, chickpea, lentil, maize, sorghum, pearl millet, wheat (Sinclair 2017) and rice (Ohsumi et al. 2008). Usually the range of VPD when partial stomata closure is initiated in VPD-sensitive genotypes is 1.5–2.5 kPa. As discussed later, such water conservation is likely to be beneficial in many grain-cropping environments. Of course, partial stomatal closure also results in a decrease in CO_2 assimilation rate that can adversely impact seasonal growth if late-season water deficit fails to develop.

Significantly, commercial cultivars are available that express the limited-transpiration trait. In maize, Corteva markets the AQUAmax line of hybrids that express the limited-transpiration trait for increased yield in the dryland regions of the US (Gaffney et al. 2015). In soybean, cultivar USDA-N8002, which also expresses the limited transpiration trait, has been released for dryland production (Carter et al. 2016). Measurements of possible partial stomata closure in 23 cultivars of Australian wheat showed that the trait was expressed in all tested cultivars with the VPD break-point occurring in the range of 1.9–2.3 kPa (Schoppach et al. 2017). The consistency in expression of the trait among all of the tested wheat cultivars was suggested to be a consequence of the development of these cultivars for production in Australian dryland conditions.

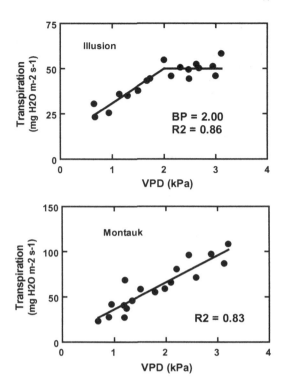

Fig. 7.2 Graph of transpiration rate versus vapor pressure deficit (VPD) for two cultivars of sweet corn (Jafarikouhini et al. 2020). The top panel is cultivar 'Illusion', which expressed limited-transpiration above a VPD breakpoint (BP) of 2.0 kPa. The bottom panel is cultivar 'Montauk', which expresses the 'classical' linear response of transpiration rate to VPD

7.4 Simulation Analysis

The potential yield advantage of partial stomata closure at elevated VPD will, of course, vary geographically depending on environmental conditions. Under late-season well-watered conditions, early-season partial stomata closure under elevated VPD will result in a negative impact on carbon accumulation by the crop. On the other hand, water conservation as a result of partial stomata closure will be advantageous if water-deficit conditions develop during the later stages of crop growth. The probability of yield increase due to the limited-transpiration trait has been explored using simulation analyses of several crop species (Sinclair 2017). For example, in an analysis of maize yield response in the US Corn Belt, Messina et al. (2015) found that the VPD water-conservation trait expressed by the AQUAmax hybrids would be especially beneficial in the western regions (Fig. 7.3).

The above discussion on crop yield, water use, and the ratio between the two help define the processes that impact the goal of 'more crop per drop'. While the phenomenological expression (Eq. 7.1) indicates that the WUE term could be a useful approach to crop improvement, it turns out that studies for improved WUE generally fail to recognize the importance of variable VPD and the temporal dynamics occurring over individual days and through the growing season. As an alternative, the skeptic can turn to a mechanistic derivation of yield as a function of leaf gas exchange

Fig. 7.3 Maps showing
simulated yield difference
(g m^{-2}) between a hybrid
with initiation of partial
stomatal closure at 2 kPa
versus a hybrid with no
stomata closure in response
to vapor pressure deficit
(Messina et al. 2015). The
three maps show the yield
difference at three levels of
exceedance probability of **a**
25% (wet season), **b** 50%
(median season), and **c** 75%
(dry season)

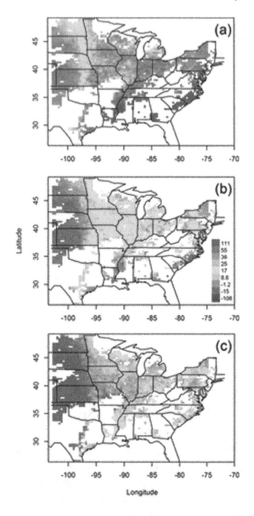

as defined by physical and physiological parameters (Eq. 7.2). Importantly, the mechanistic derivation shows that partial stomata closure under elevated VPD has the possibility of increasing soil water availability during late-season crop growth. Such water conservation can contribute to sustained crop physiological activity during the critical period of reproductive development if soil water availability is inadequate. The possibility of yield increase has been supported by simulation analysis and the development of commercial cultivars expressing the limited-transpiration trait.

References

de Wit CT (1958) Transpiration and crop yields. Institute of Biological Chemistry Research, Field Crops and Herbage, No. 646. Wageningen, The Netherlands.

Farquhar GD, O'Leary MH, Berry JA (1982) Isotopic composition of plant carbon correlates with water-use efficiency of wheat genotypes. Aust J Plant Physiol 9:121–137

Gaffney J, Schussseler W, Loffler C, Cai W, Paszkiewicz S, Messina CD, Groeteke J, Keaschall J, Cooper M (2015) Industry-scale evaluation of maize hybrids selected for increased yield in drought-prone conditions of the US Corn Belt. Crop Sci 55:1608–1618

Jafarikouhini N, Pradhan D, Sinclair TR (2020) Basis of limited-transpiration rate under elevated vapor pressure deficit and high temperature among sweet corn cultivars. Env Exp Bot 179:104205

Messina CD, Sinclair TR, Hammer GL, Curan D, Thompson J, Oler Z, Gho C, Cooper M (2015) Limited-transpiration trait may increase maize drought tolerance in the US Corn Belt. Agron J 107:1978–1986

Ohsumi A, Hamasaki A, Nakagawa H, Homma K, Horie T, Shiraiwa (2008) Response of leaf photosynthesis to vapor pressure differences in rice (Oryza sativa L.) varieties in relation to stomatal and leaf internal conductance. Plant Prod Sci 11:184–191

Passioura JB (1977) Grain yield, harvest index and water use of wheat. J Aust Inst Agric Sci 45:117–121

Rizza F, Ghashghaie J, Meyer S, Matteu L, Mastrangelo AM, Badeck F-W (2012) Constitutive differences in water use efficiency between two duru wheat cultivars. Field Crops Res 125:49–60

Schoppach R, Fleury D, Sinclair TR, Sadok W (2017) Transpiration sensitivity to evaporative demand across 120 years of breeding of Australian wheat cultivars. J Agro Crop Sci 203:219–226

Sinclair TR (2012) Is transpiration efficiency a viable plant trait in breeding for crop improvement? Func Plant Biol 39:359–365

Sinclair TR (ed) (2017) Water-conservation traits to increase crop yields in water-deficit environments. Springer Briefs in Environmental Science, Switzerland

Sinclair TR, Bennett JM, Muchow RC (1990) Relative insensitivity of grain yield and biomass accumulation to drought in field-grown maize. Crop Sci 30:690–693

Spaeth SC, Randall HC, Sinclair TR, Vendeland JS (1984) Stability of soybean harvest index. Agron J 76:482–486

Tanner CB, Sinclair TR (1983) Efficient water use in crop production: research or re-search? In: Taylor HM, Jordan WR, Sinclair TR (eds) Limitations to efficient water use in crop production. Am Soc Agron, Madison, WI, pp 1–27

Vadez V, Kholova J, Medina S, Kakkera A, Anderberg H (2014) Transpiration efficiency: new insights into an old story. J Exp Bot 65:6141–6153

Wong SC, Cowan IR, Farquhar GD (1979) Stomatal conductance correlates with photosynthetic capacity. Nature 282:424–426

Chapter 8
Transpiration Prediction

Precipitation, or more specifically the lack of precipitation, is expected to be a major consequence of climate change in many cropping areas. In addition to altered precipitation patterns, climate change is predicted to result in greater atmospheric vapor pressure deficit (Ficklin and Novick 2017). Both of these climate changes have now been documented to be occurring and resulting in more-arid conditions for plant growth, including in many crop production areas. To counterbalance the anticipated increasing water deficits, alterations in crop management need to be developed. The management alterations require estimates of the anticipated transpiration rate by crops to allow tracking of soil water content and its potential impact on crop growth and yield.

8.1 Energy Balance Approach

The basis of one very popular approach to estimating crop water loss calculates the energy balance at the crop surface (Allen et al. 2005). The fundamental equation is based on conservation of energy flux density among radiation, sensible heat, latent heat (i.e. water flux density), heating of the soil, and photosynthesis. The energy balance is dominated by radiation, and the sensible and latent heat flux densities so the other terms are usually ignored. This simplified equation offers a fairly accurate description for estimating **instantaneous** crop water loss rate.

Caution: Time

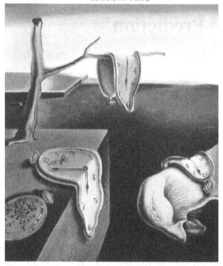

Critical issues arise, however, when applying the energy balance equation to estimating water loss rates over **integrated time** periods of several hours or an entire day or longer. While in practice the energy balance equation has been readily applied in the calculation of water loss over extended-time periods, it is being done without critical consideration of the temporal dynamics of the input variables. This becomes apparent when the energy balance equation is actually written in integral format to define the water loss at a surface over an extended time period of interest.

$$E = \int \left[R_n \Delta + \gamma f(u) VPD\right] / \left[\Delta + \gamma\right] dt \tag{8.1}$$

where

E   water loss rate
t   time
R_n net radiation
Δ slope of change in vapor pressure with respect to temperature
γ psychrometric constant
f(u) aerodynamic conductance function dependent on wind speed
VPD vapor pressure deficit

The interacting temporal dynamics of variables within the integral must be taken into account when applying the energy balance equation for circumstances other than very short time intervals.

Harold Penman (1948) in his classic paper applied the energy balance equation without specifically accounting for integration over time. That is, he applied the energy balance equation directly to time durations of a day to months. This resulted in a challenge in how to determine appropriate values for non-instantaneous time periods. Net radiation, which Penman (1948) calculated from the ratio of actual to possible hours of sunshine for the period being studied, was simply summed over the time period being considered. Determining values for the weather-dependent variables f(u) and VPD were more difficult. Penman's solution for a daily value of f(u) was obtained by an empirical equation based on measured miles day^{-1}. Hence, variation in wind speed through the 24 h day was not taken into account even though the generally low wind speed hours of the night when crop water loss is low was simply included to obtain an average for the total. Calculations for periods longer than a single day were simply based on an average daily wind speed for the entire period of interest.

Penman calculated average daily VPD from temperature measurements. The atmospheric vapor pressure was calculated based on a single daily measure of atmospheric dew point, which Penman noted was not always "adequate". Since Penman's initial focus was evaporation from an open water surface, daily saturation vapor pressure was calculated by measuring water surface temperature at 4-h intervals and then using an average temperature of the six values to calculate saturation vapor pressure, even though saturation vapor pressure is not linearly dependent on temperature.

For a plant surface, which in Penman's case was a turf grass, variations in plant water loss through the 24 h daily cycle due to changing influences of VPD and constraints by stomata vapor conductance were simply ignored. That is, canopy water loss was treated as occurring over a 24 h day length, which is clearly not the case considering that during the night both VPD and stomata conductance are generally quite low. Calculations based on a 24 h perspective, particularly in obtaining a 24 h average VPD, continues to be an error in many approaches in estimating crop water loss.

Using the above assumptions and empiricisms, Penman (1948) used the formulation of the instantaneous energy balance equation to calculate evaporation from an open-water surface. He then compared the calculated estimates for water loss from an open-water surface with measured water loss from a turf grass. Penman found that the open-water estimate resulted in an overestimation of grass water loss in all cases. To estimate water loss for the turf grass that he studied in the United Kingdom, Penman suggested simply decreasing the open-water estimate by approximately 0.6 in the winter and by 0.8 in the summer. The introduction of this variable coefficient in the multiplication of the open-water evaporation added another challenge in the integration of water loss through the growing season. Not surprisingly, Penman (1948) cautioned about the "empirical aspects" of his approach and "until these are removed there must always be some doubt about the possibility of successful translation in space and time of the [energy balance] formulae".

As an improvement in using the energy balance equation to estimate crop transpiration, John Monteith (1964) suggested the inclusion of a 'canopy conductance' into the energy balance equation. This canopy conductance was conceived as a term related to the restrictions on vapor diffusion as a result of plant stomata and boundary layer diffusion resistances. However, evaluation of canopy conductance has remained unresolved due the complexity in the variability within the leaf canopy in stomata vapor conductance and the location within the canopy of the main sources of transpired water. Stomata conductance and transpiration rate can be quite variable through a day and over a series of days in response to light conditions, vapor pressure deficit, and plant nutrient and water status. No clear method has been developed for making an *a priori* estimate for Monteith's canopy conductance integrated over time.

While the Penman-Monteith approach seems hypothetically appealing, it is essentially impossible to apply it directly to actual field conditions. A rather curious solution to this difficulty has evolved, however, allowing retention in the use of the Penman-Monteith equation. In this approach, the energy balance equation is used to calculate a 'reference value' for crop water loss by simply setting fixed values for several variables in the energy balance equation based on rather crude assumptions (Allen et al. 2005). The aerodynamic conductance based on wind speed involves simply assuming aerodynamic conductance is a function of canopy height. Canopy resistance, the inverse of Monteith's canopy conductance, is determined using very simplistic assumptions about canopy transpiration, and the value of canopy resistance is ultimately simply set to a constant of 70 s m^{-1}. Further, the original error of using a 24 h day VPD for transpiration instead of a daily weighted VPD for transpiration was carried forward. In spite of the dubious nature of the calculation of the reference transpiration, the United Nations Food and Agriculture Organization (UN-FAO) certified this approach "as the sole method for determining reference evapotranspiration" (Allen et al. 2005). (By the way, the use of the word 'evapotranspiration' by Allen et al. (2005) is a misnomer because the canopy conductance modification of Monteith was clearly in the context of defining canopy transpiration rate.)

"Actual" crop water loss is calculated from the output of the reference equation approach by multiplying it by an unknown and unknowable variable "crop coefficient". Tables and figures have been generated to aid in guessing values for what might be appropriate crop coefficients under any particular condition (Allen et al. 1998). The value of this empirical crop coefficient varies from near zero to values greater than 1.0. Various suggestions are sometimes offered to help in selecting what might be an appropriate guestimate of the crop coefficient under changing circumstances.

Given the extent of the assumptions and approximations of the UN-FAO approach in estimating transpiration rate, it should be no surprise, at least to the skeptic, that the approach is inherently fallible. For example, in an attempt to determine the crop coefficient for four, well-watered grass species, which were expected to have a fairly constant, common crop coefficient value, we (Wherley et al. 2015) found that its value was far from constant. The experimentally determined values for the crop coefficient varied widely from 0.17 to 1.00 (Fig. 8.1) with significant differences also existing among these very similar grass species.

Given the assumptions, approximations, and uncertainty in the use of the energy balance approach describe above, it is surprising that Penman's (1948) alternate and simpler approach of a sink-strength model has been virtually ignored. The sink-strength model calculated evaporation simply as a function of a surface conductance multiplied by the vapor pressure deficit. Conductance was estimated by Penman based on the miles day^{-1} wind speed and vapor pressure deficit was estimated as a 24 h value. In fact, Penman pointed out that the predictive capability of evaporation

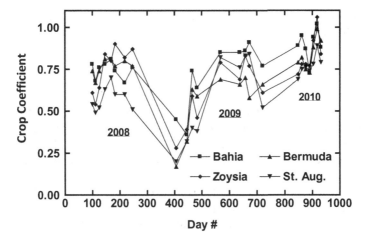

Fig. 8.1 Crop coefficient calculated from energy balance for 2–4 days periods over three years for four turf grasses (Wherley et al. 2015)

from an open-water surface by the energy-balance and by the sink-strength model were approximately equal.

8.2 Plant Gas Exchange Approach

The importance of variation in vapor pressure deficit in the sink-strength model of Penman (1948) to estimate water loss is consistent with the prominence of VPD in the derivation of canopy yield and transpiration by Tanner and Sinclair (1983) that was presented in the previous chapter. Equation (8.2) in the previous chapter can be rearranged to calculate transpiration amount from the following equation.

$$T = \int (G \cdot VPD)dt \Big/ \left(HI \cdot k \cdot \int dt\right), \tag{8.2}$$

where

T	transpiration amount
t	time
G	plant mass accumulation amount
VPD	vapor pressure deficit (Pa)
HI	harvest index
k	mechanistic coefficient explicitly defined by crop physical and physiological characteristics (Pa).

If transpiration rate is to be estimated for a single day (T_{day}) based on daily growth (G_{day}) using Eq. (8.2), then an option could be to base the calculation on an 'effective VPD' (VPD_{eff}) for that day. The value of VPD_{eff} needs to reflect the weighted VPD exposure during the integrated period of active crop transpiration. In this case, Eq. (8.2) simplifies to the following equation.

$$T_{day} = G_{day} \cdot VPD_{eff} / k \tag{8.3}$$

The value of VPD_{eff} was suggested by Tanner and Sinclair (1983) to be "three quarters" of the value between the daily maximum difference in saturated vapor pressure at maximum temperature and actual atmospheric vapor pressure. The actual atmospheric vapor pressure can be estimated by assuming daily minimum temperature decreases to approximately the dew point on each night. Abbate et al. (2004) did an extensive evaluation of the fraction of maximum daily maximum VPD that appropriately represented VPD_{eff} during the period of wheat production in Argentina. They found that the fraction was consistent with the suggestion of Tanner and Sinclair (1983) in obtaining a value of 0.72.

The major challenge in applying Eq. (8.3) appears to be in the determination of the value of G_{day}, daily plant mass accumulation. However, this challenge can be solved by defining G_{day} as being equal to radiation use efficiency (RUE) multiplied by the amount of daily intercepted solar radiation. Intercepted solar radiation is, in turn, defined by the solar radiation incident to the crop canopy (I_o) multiplied by the fraction of incident radiation intercepted by the canopy $(1 - \exp[-g \cdot LAI / \sin \beta])$, where g = leaf shadow projection, LAI = leaf area index, and β = sun angle. The crop canopy LAI can be tracked day-to-day based on cumulative temperature after sowing or after plant emergence from the soil. Since the daily value of $g/\sin\beta$ is approximately 0.7 for latitudes around 30° and leaf angle about 50° (Sinclair 2006), the calculation of radiation interception is approximately $(1 - \exp[-0.7 \cdot LAI])$. As discussed in Chap. 3, values of RUE are fairly well established for unstressed crops. Therefore, transpiration rate can be obtained by substituting into Eq. (8.3) the definition of G_{day}.

$$T_{day} = RUE \cdot I_o \cdot (1 - \exp[-0.7 \cdot LAI]) \cdot VPD_{eff} / k, \tag{8.4}$$

If transpiration is to be calculated for a closed crop canopy, then Eq. (8.4) can be simplified since high LAI canopies intercept virtually all incident radiation. Therefore Eq. (8.4) becomes

$$T_{day} = RUE \cdot I_o \cdot VPD_{eff} / k \tag{8.5}$$

Hence, transpiration can be calculated using Eq. (8.5) without dubious assumptions regarding estimating unresolved empirical values for any variable.

Wherley et al. (2015) used Eq. (8.5) to predict the water loss of the four turf grass species discussed above. The value of RUE for these grasses was set equal to 1.0 g MJ^{-1} solar radiation based on published observations (Kiniry et al. 2007; Cristiano et al. 2012). As discussed in Chap. 7, the value of k for these C4 grasses is

9 Pa. The weighted value for daily VPD_{eff} was set equal to the maximum daily VPD multiplied by 0.75 (Tanner and Sinclair 1983).

The results of the predicted transpiration rate vs. observed transpiration rate for each 2–4 days observation period and each grass measured periodically during three years are graphed in Fig. 8.2 (Sinclair et al. 2014). The predicted vs. observed transpiration rates were well correlated for each grass by linear regressions through the origin (P < 0.0001 in each case). The slope of the regression in each case was somewhat greater than 1.0, which was concluded to be a result of a slight overestimation of the amount of canopy radiation interception.

The analysis of transpiration of the well-watered turf grasses shown in Fig. 8.2 was based on a non-stressed leaf canopy, which meant RUE was well defined. However, stress conditions associated with loss in leaf photosynthesis activity results in decreased values of RUE (Fig. 3.1), and consequently the value of growth (G_{day}). Fortunately, functions have been developed to describe the impact on RUE of various stresses such as cool temperature (Andrade et al. 1993) and leaf nitrogen decrease (Cabrera-Bosquet et al. 2016). Soil water deficit can have especially negative impacts on leaf gas exchange, and hence result in decreased RUE (Sinclair 2017). In the case of soil water deficits, the deficits can be tracked by rather simple soil water balance models to obtain daily soil water content. Based on soil water content expressed as fraction transpirable soil water (FTSW), the value of RUE can be decreased using a response function such as presented in Fig. 6.1. While stresses can complicate

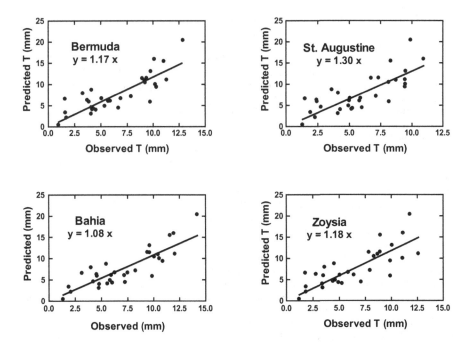

Fig. 8.2 Graphs for four turf grasses of predicted water loss calculated using Eq. (8.5) versus measured water loss over 2–4 days measurement periods during three years of observation (Sinclair et al. 2014)

calculations somewhat, the degree of stress can be tracked and adjustments made in daily RUE values as the input to Eq. (8.5). Hence, the output of Eq. (8.5) allows a prediction of canopy transpiration rate from well-defined variables that relate directly to the regulation of leaf gas exchange and crop growth.

Consequently, the continued use of the empirical and ambiguous energy balance equation to predict transpiration of vegetative surfaces does not seem warranted. Transpiration estimation based on mechanistic expressions of plant gas exchange (Eq. (8.2)) offers a straightforward approach to calculating transpiration rate. The variables for the calculation are clearly defined and can be tracked for field conditions with rather simple models.

References

Abbate PE, Dardanelli JL, Cantarero MG, Maturano M, Melchiori RJM, Suero EE (2004) Climate and water availability effects on water-use efficiency in wheat. Crop Sci 44:474–483

Allen RG, Pereira LS, Raes D, Smith M (1998) Crop evapotranspiration—guidelines for computing crop water requirements. FAO Irrigation and Drainage Paper 56. FAO—United Nations, Rome

Allen RG, Walter IA Elliott A, Howell T, Itenfisu D, Jensen M (2005) The ASCE standardized reference evapotranspiration equation. American Society of Civil Engineering—Environmental and Water Resources Institute of Task Committee Report

Andrade FH, Uhart SA, Cirilo A (1993) Temperature affects radiation use efficiency in maize. Field Crops Res 32:17–25

Cabrera-Bosquet L, Fournier NB, Welchker C, Suard B, Tardieu F (2016) High-throughput estimation of incident light, light intercept and radiation use efficiency of thousands of plants in a phenotype platform. New Phytol 272:269–281

Cristiano PM, Posse G, DiBella CM, Boca T (2012) Influence of contrasting availabilities of water and nutrients on the radiation use efficiencies of C3 and C4 grasses. App Eco 37:323–329

Ficklin DL, Novick KA (2017) Historic and projected changes in vapor pressure deficit suggest a continental-scale drying of the United States atmosphere. J Geophy Res: Atmos 122:2061–2079

Kiniry JR, Burson BL, Evers GW, Williams JR, Sanchez H, Wade C, Featherston JW, Greenwade J (2007) Coastal bermudagrass, bahiagrass, and native range simulations in diverse sites in Texas. Agron J 99:450–461

Monteith JL (1964) Evaporation and environment. In: The State and Movement of Water in Living Organisms, 19th Symposia of the Society for Experimental Biology. Cambridge University Press, Cambridge, pp 205–214

Penman HL (1948) Natural evaporation from open water, bare soil and grass. In: Proceedings of the Royal Society of London. Series A Math Phys Sci, vol 193, pp 120–145

Sinclair TR (2006) A reminder of the limitations in using Beer's Law to estimate daily radiation interception for vegetation. Crop Sci 46:2342–2347

Sinclair TR (ed) (2017) Water conservation traits to increase crop yields in water-deficit environments. Springer Books in Environmental Sciences, Springer, Switzerland

Sinclair TR, Wherley BG, Dukes MD, Cathey SE (2014) Penman's sink strength model as an improved approach to estimating plant canopy transpiration. Agric Met 197:136–141

Tanner CB, Sinclair TR (1983) Efficient water use in crop production: research or re-search? In: Taylor HM, Jordan WR, Sinclair TR (eds) Limitations to efficient water use in crop production. American Society of Agronomy, Madison, WI, pp 1–27

Wherley B, Dukes MD, Cathey S, Miller G, Sinclair T (2015) Consumptive water use and crop coefficients for warm-season turfgrass species in the Southeastern United States. Agric Water Manag 156:10–18

Chapter 9
Unconfirmed Field Observations (UFOs)

A cornerstone of scientific investigation is open-mindedness and critical evaluation of evidence, i.e. skeptical analysis. For crop science, an essential source of evidence is almost always experiments that are well-structured and unbiased. To gain insight about whether a specific treatment has improved crop performance, the results of a treatment need to be compared with a 'standard' treatment, sometimes labelled as a control. Attempts are made to have all variables equal between the treatment(s) and the standard except for the specific treatment(s) of interest. In this way, the results can be interpreted as solely a result of the altered treatment(s). The interpretation of the experimental results is often aided by statistical comparison, but the experience and insight of the investigator is the essential component in developing conclusions.

Caution: UFO

Unconfirmed Field Observation (UFO) ?

Crop science research is challenging, however, because results can be impacted by so many variables that it is virtually impossible to have all variables controlled. This is especially true since the final conclusion about the success of a treatment to increase crop yield is expected to be applied to a range of practical field conditions. The research can be tedious, long-lasting, and finally confusing. One approach found attractive by some is to 'short-cut' the experimental research involving the usual scientific and agronomic practices that provide evidence for objective interpretation.

© The Author(s), under exclusive license to Springer Nature Switzerland AG 2022   59
T. Sinclair and T. W. Rufty, *Bringing Skepticism to Crop Science*, SpringerBriefs
in Agriculture, https://doi.org/10.1007/978-3-031-14414-1_9

Instead, data are obtained from farmer yield contests or from abbreviated field trials. We (Sinclair and Cassman 2004) labelled these short-cut approaches as Unconfirmed Field Observations (UFOs).

9.1 Yield Contests

As presented by Shermer (1997) and discussed in Chap. 1, three possible reasons why UFO observations have been used is they often match the need for *Credo consolans*, immediate gratification, and simplicity. (For Shermer, these reasons were applied to understanding the reasons for recurrent reports of space UFOs.) Yields from farmers' yield contests are sometimes reported in the scientific literature to satisfy these needs. One of the more recent examples of such a paper was an attempt to analyze evapotranspiration amounts based on contest reports for high-yield maize (Basso and Ritchie 2018). A difficulty is that the contest winning yields may simply be greater than the solar energy available to support the synthetic processes required to achieve such yields.

An evaluation of the possible realism of a farmer reported yield can be done by a fairly simple calculation of maximum grain yield (Y_{max}) based on input solar energy. The following simple equation can be used based on the incident solar radiation over a growing seasons.

$$Y_{max} = HI \int (I \cdot RUE)dt \tag{9.1}$$

where

HI   harvest index
I   daily intercepted solar radiation
RUE   radiation use efficiency
t   time

Assuming high values of the input variables for maize (*Zea mays* L.) of HI = 0.5, $I = 20$ MJ m^{-2} day^{-1}, RUE = 1.8 g MJ^{-1} and a fully productive growing season of 110 days, the value of Y_{max} is 1980 g m^{-2}. Accounting for the standard 15.5% seed moisture allowed in marketed maize grain, the maximum yield is calculated to be 2343 g m^{-2} (348 Bu A^{-1}). While this grain yield would be enthusiastically welcomed by nearly all farmers, the 2019 US maize-yield contest winner reported a yield of 616 Bu A^{-1} (!), or almost double what is possible based simply on a very high estimate for solar energy input. The uncontrolled nature of these contest UFO observations that ignore the usual scientific standards for field research is a major warning against using such reports in any scientific analysis or interpretation.

9.2   Gene Transformations

In recent years, however, a somewhat more insidious UFO has invaded the literature. The sources of many of these reports are from scientists working at the molecular level who are interested in demonstrating the impact of a gene transformation on whole plant performance. A gene transformation is made, lines are selected that positively express the desired molecular change, and plants are grown out under controlled environmental conditions or in an isolated field trial to show growth response. Erroneously, positive growth responses measured in various ways from these tests are often referred to as yield increase. This conclusion ignores the fact that the criteria for claiming a 'yield increase' are more demanding than the rather naïve, simple comparison with the original wild-type genotype, which is often the basis for these studies.

An essential, primary criterion for claiming yield increase is that experiments need to be done under field conditions. While growth chambers and greenhouses offer excellent facilities to explore molecular and physiological behavior, and to conduct preliminary assessments of growth adjustments, the results do not necessarily relate to field responses. Confounding factors in controlled environment comparisons can include low light levels, high fractions of diffuse light, square-wave temperature regulation, no or poor regulation of vapor pressure deficit, a potting media that is unlike field soil particularly in unrealistic soil water release, and limited pot volume resulting in pot binding. Ultimately, yield testing must be done under realistic field conditions.

An associated criterion to the above is that comparisons of yield capability cannot be inferred from individual plants grown either in controlled-environment facilities or in the field. Yield in the field for most crops is produced by a community of plants forming a closed canopy or a nearly closed canopy. The environment of an individual plant grown by itself or plants grown in small, unborder plots are substantially different from that of a closed canopy, and the limiting growth processes are different. It cannot be assumed that the community of plants that individually have strong growth will necessarily translate into a large positive response for plants grown in field production.

Field tests must include all the usual cautions for assessing yield. The first concern is to select field sites within which there is a uniformity in soil and topography. Then, when installing a field test it is necessary to use row widths and plant densities that are consistent with the commercial production of the crop being studied. Individual plots need to be sufficiently large so there is border area on all sides of the plot and only the central bordered area of the plot is harvested to measure yield. The material harvested needs to be based on some defined ground area, not by number of plants. The area to be harvested must be sufficiently large to account for the plant-to-plant variability that may occur in a test plot. Plot replication in harvesting is required, which means plots need to be arranged in an appropriate statistical design. Finally, care is required in handling harvested material, which usually includes drying harvested material to some common, accepted tissue moisture content.

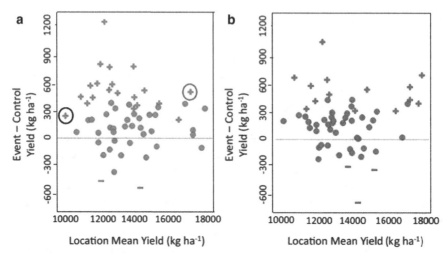

Fig. 9.1 Yield difference between hybrids derived from two transformed elite inbreds (A. P202216 and B. DP382118) with the *zmm28* gene labelled as "event" and wild-type hybrids derived from untransformed elite inbred lines (Wu et al. 2019) labelled as "control". Those environments in which the hybrids from the transformed inbred having greater yield with p < 0.05 than the untransformed inbred are shown in a + symbol. Those cases with decreased yield are shown in a – symbol. The two cases encircled in black and red highlight that the range of positive response observed over the full range of tested yields

An important feature of field yield trails is the number of 'environments' in which a field test must be done to resolve the question of yield increase. Differing environments can be achieved either by using several locations or several growing season, or preferably both. Sinclair et al. (2020) examined the output from several simulation studies to assess the number of environments from which results are required to obtain stability in predicted yields. They concluded that up to 30 environments may be required to capture the full impact of weather variability on crop yield. The basic criterion is to do yield tests in a sufficient number of environments to reflect the range of conditions in which the derived genotype is to be grown.

Finally, a major issue in evaluating field test results is to determine what genetic material is to be compared. In many recent molecular genetic papers, comparisons are made between transformed lines and their source 'wild type' line. However, this comparison does not meet the usual agronomic standard in resolving yield increase in which comparisons need to be made with existing, high-yielding commercial cultivars. The necessity of the comparison with a commercial cultivar is apparent when it is recognized that the wild-type germplasm may be weak-yielding genetic material. Such weak plants are often used in transformation studies because they are more vulnerable to transformation techniques. Hence, yield tests to compare the impact of genetic modification with the high-yielding commercial lines must wait until gene introgression into existing high-yield material can be done.

An example of a study that successfully took into account the UFO concerns discussed above was published by Wu et al. (2019). This was a major study by Corteva Agriscience involving 29 investigators. The objective of the study was to assess the yield impact of increasing the expression of the transcription factor *zmm28* gene, which seemed to have several positive benefits for maize growth. The gene was inserted into a semi-elite inbred line and then introgressed into two elite inbred lines. These two elite lines were crossed with high-yielding inbred lines to obtain a total of 48 hybrids. Yield tests were done over four years involving 24 locations in North America and Chile. Two to three replicates were established in each test in 4-row plots that were 4.5–5.5 m long. The two center rows were harvested and the grain dried to 15% moisture. The yields of the transgenic hybrids varied relative to the wild-type hybrids derived from each of the elite inbred parents (Fig. 9.1). The raw yields of the transgenic hybrids derived from the two elite inbred lines was greater than the wild-type hybrid in 78 and 79% of the comparisons. Overall, these results showed in field tests a positive yield response to the transformation but, not surprisingly, the results were variable. This variation is an important metric in marketing a commercial product to farmers.

It is essential for the skeptic to be fully aware of the test conditions on which claims of yield increase are based. Yield results from field measurements are the ultimate arbiter in documenting yield improvements. Unfortunately, UFOs are now a part of the scientific literature. Whether it is uncontrolled yield results from farmer contests or inadequate tests of transformed genetic lines, these results need to be set aside until well controlled, scientifically adequate field trials are done.

References

Basso B, Ritchie JT (2018) Evapotranspiration in high-yielding maize and under increased vapor pressured deficit in the US Midwest. Agric Environ Lett 3:170039

Shermer M (1997) Why people believe in weird things. WH Freeman and Company, New York

Sinclair TR, Cassman KG (2004) Agronomic UFOs. Field Crops Res 88:9–10

Sinclair TR, Sotani A, Marrou H, Ghanem M, Vadez V (2020) Geospatial assessment for crop physiological and management improvements with examples using the simple simulation model. Crop Sci 60:700–709

Wu J, Lawit SJ, Weers B, Sun J, Mongar N et al (2019) Overexpression of zmm28 increases maize grain yield in the field. Proc Nat Acad Sci 116:23850–23858

Final Comments

Dear Reader

First, thank you for taking the time and making the effort to read at least part of this book. Tom Rufty and I enjoyed working together to prepare "Bringing Skepticism to Crop Science". We hope you found your reading interesting, and maybe even thought-provoking.

The idea for the book was to illustrate that skepticism is an integral part of science. Without skepticism, a cornerstone of science is missing. Biological truths are elusive, requiring step-wise advances in knowledge, with steps often being only loosely assembled. An open-minded and critical evaluation of evidence is necessary to judge individual hypotheses. Without skepticism it also seems to us that biological science becomes very bland and a person is missing an essential part of scientific inquiry and its pursuit of truth. Penetrating questions are key. Was evidence collected without bias? Were experiments done using protocols that considered assumptions and errors? Was the evidence evaluated in the broad scope of possibilities and not just to support existing hypotheses? And finally, after reviewing evidence from a skeptical viewpoint are new hypotheses and ideas possible?

In this book, we attempted to indicate that some of the readily accepted ideas in crop science are vulnerable to alternate concepts once the prejudiced of past teachings and literature are subjected to a skeptical review. We have attempted to re-examine some of the basic hypotheses in crop science that have been retained for decades even though the fundamental support for them remains tenuous. Alternate interpretations of the evidence are possible, and at least for us, the alternative interpretations make more sense in understanding crop science and in anticipating the next important research topics that should be investigated.

We hope, however, the obvious message to you from this book is "be skeptical". Skepticism about conclusions discussed in this book applies not only to existing ideas but maybe especially to the alternative hypotheses we proposed! Do not accept anything presented in this book, or elsewhere, without a skeptical attitude. All ideas presented here should be approached with an open mind and critical evaluations to make your own decisions about what makes sense to you. Enjoy the skeptical perspective as a fundamental component in creating new evidence, evaluating all evidence, and arguing for new conclusions—to us this is the fun of science!

Tom Sinclair

Printed in the United States
by Baker & Taylor Publisher Services